DELAY AGEING

Healthy to 100

Colin Rose

First published in the United Kingdom in 2020

Accelerated Learning Systems Ltd
12 The Vale, Southern Rd, Aylesbury, HP19 9EW.

acceleratedlearning.com

ISBN 9780905553702 (paperback)
ISBN 9780905553719 (ebook)

ABOUT THE AUTHOR

Colin Rose graduated from The London School of Economics in 1961.

He is a Fellow of The Royal Society of Arts and a Senior Associate of the Royal Society of Medicine. He has written on health, nutrition, science and learning for over 35 years and edited Dr Paul Clayton's best-selling book *Health Defence*.

Colin has also been an international educational advisor. Author of *Accelerated Learning* and *Accelerated Learning for the 21st Century*, which were translated into 12 languages, Colin was a main contributor to *EduScience* – a programme funded by the European Union which aimed to enhance the teaching and learning of science in schools in Europe.

He was a founder member of the UK's *Campaign for Learning* – a collaboration between government and industry.

Colin founded Accelerated Learning Systems in 1984 and Uni-Vite Healthcare in 1986. Delay Ageing is his tenth book.

He says:

"Although I am not a biologist, a geneticist, a physician or a gerontologist, I believe I do know how to interpret scientific research and make it accessible for a general audience.

"That's what I did many years ago, when I took the then latest research on the brain, learning and memory to develop and write *Accelerated Learning*.

"So, this book combines both my interests – education and health. Its purpose is to make some very important research, which not many people would read in its original form, both understandable and actionable.

"I leave you to judge if I have succeeded."

EDITOR'S NOTE

The book uses British spelling in the main text. However, direct quotes, titles of books and papers, and names of institutions and organisations are generally reproduced in their original published spelling. Hence, the spellings "ageing" and "aging" both appear.

DELAY AGEING

by Colin Rose

continued overleaf

INTRODUCTION

WHY WE NEED TO DELAY AGEING

What if getting older didn't mean getting ill?

Most people accept that growing older means suffering illness –
less energy, a risk of cardiovascular disease, brain disorders, cancer,
aching joints and increasing frailty. Most people shrug and say: "It's
just the way life is".

You'll be pleased to know that most people are wrong.

Researchers at universities like Harvard Medical School,
Cambridge, Oxford, University College London, UC Berkeley
have all – over the last decade – come to another conclusion.

It's best summed up by Professor Linda Partridge, Director of
the Institute of Healthy Ageing at University College London:

> *"Ageing is a malleable process."*

You can directly influence how fast, or slow, you age.

David Gems is Professor of the Biology of Ageing at University
College London and Deputy Director of the Institute of Healthy
Ageing. He agrees:

> *"If ageing is delayed in humans you would have a reduction in
> most or all ageing related illnesses – cancer, dementia such as
> Alzheimer's, cardiovascular disease, type 2 diabetes, blindness,
> osteoporosis."*

Join me in a biological detective journey into the new science of
healthy ageing. You need no prior knowledge but be prepared for

1

surprises – because it's not just a simple matter of eat your greens and move more.

What you'll learn can transform your health – or your parents' health – as it has mine. You'll discover exactly what Silicon Valley billionaires are investing fortunes to learn.

Of course, they are planning to develop patentable drugs, wearable health monitors or high-tech procedures like gene editing or stem cell transplants. Because they know that if we can delay ageing, we can delay the onset of 'age related' diseases and the market for their products will be enormous.

In contrast, every single age-delaying strategy we will explore will involve natural solutions that you can start tomorrow – whatever your age. Because you will discover changes you can make in your 50s, 60s 70s, even 80s that will result in a healthier, longer life.

It's never too late.

Health extension rather than life extension

The potential impact on your personal quality of life in your later years is transformative, but extending your healthy lifespan also has a significant economic benefit.

Richard Faragher is a Professor of Biological Gerontology at the University of Brighton, England and Chair of the British Society for Research on Ageing. In a recent video he points out that over half of people over the age of 65 will ultimately spend over £20,000 of their own money on social care, with 10% of them spending over £100,000. Money they had hoped to leave to their children.

The economic savings of delaying ageing for society would be in the trillions – since an estimated 40% of the National Health budgets in the UK and USA are spent on the years when age-related diseases have surfaced.

The aim is NOT to pursue a delusional 'fountain of youth fantasy' of living to 150. That would be planet destructive. The aim IS to extend the years when you live fit, happy and well with your family – though some life extension is likely.

There's no point in living longer unless it's fun to be alive.

The other aim is to be able to continue to contribute productively to society. Indeed, a large cohort of fit and well older people working, for example, with charities, contributing economically and on projects to improve local environments, could greatly benefit society.

Indeed, we MUST engage with healthy ageing. Because if we don't improve people's health in later years, the cost to society in medical bills will be an unsustainable burden for the decreasing numbers of younger full time employed.

That's because our current model of healthcare is flawed. The principal focus is on cure, not prevention. Wait until a disease has surfaced and zap it with a 'magic bullet' drug. In fact, it's worse than that. Most of the time patients are not cured, but merely enabled to live with the disease rather than die of it.

Fortunately, we now know how to restore tissue and organ function to a younger state. How to separate biological from chronological ageing. How to increase 'health span' rather than just life span.

The scientists measure it as Quality Adjusted Life Years (QALYs). I prefer more life in your years, rather than more years in your life.

Cutting edge science

There are well over 100 university and national centres of ageing research around the world – many individual scientists are listed in Appendix 1.

These researchers are generally agreed that there are nine, universal *'Hallmarks of Aging'* – originally identified in a much-cited paper in *Cell Journal* 2013 authored by Carlos López-Otín, Linda Partridge and others[1].

These hallmarks are biological processes, common to all of us, that underly ageing and which – if they are not counteracted – will inevitably lead to 'age related diseases'.

I have reviewed hundreds of studies and spoken to health researchers and the science is clear: each one of these markers, or 'hallmarks', of ageing can be slowed, delayed, or in some cases stopped or even reversed. You'll even meet a researcher whose test subjects actually aged backwards!

The result is to increase the years you stay healthy.

The nine **Hallmarks of Ageing** are as follows.

1 Damage to DNA accumulates

This leads to 'gene instability', to mutations and loss of cell function. DNA damage is central to ageing, cancer and deeply implicated in both Alzheimer's and heart disease. But it is possible to boost your level of cell repair.

2 Cells become 'senescent'

When a cell ages and can no longer function properly, it is normally replaced with new healthy cells. As they age, however, some cells deteriorate but do not completely die.

These senescent cells hang around like 'zombies' and pour out toxins that cause inflammation. Inflammation promotes ageing, and is a key driver of atherosclerosis, heart disease, diabetes, dementia and arthritis, creating a condition where cancer cells can spread. The good news: researchers have discovered how to help clear away these 'zombie cells'.

3 Mitochondria become dysfunctional

Mitochondria are the tiny power plants in almost every cell. Dysfunctional mitochondria lead to loss of energy, muscle weakness, fatigue and cognitive problems. However, there are specific foods and nutrients that can boost your mitochondrial repair.

4 Beneficial genes are switched off, harmful genes are on

Your genes are fixed, but the way they are 'expressed' – turned on or off – is something you have some significant control over. Scientists call this 'epigenetic change'. We will see how certain foods and lifestyles can turn on genes that contribute to health and turn genes off that lead to disease.

5 Stem cells become exhausted

Stem cells can develop into different cell types, from brain cells to muscle cells as needed. But the body has a limited number of adult stem cells and the number declines with age. It is possible, however, to slow down the rate of stem cell decline.

6 Cells fail to communicate properly

Cells need to 'talk' to each other and to sense each other's boundaries otherwise disease and especially cancer can develop. Fortunately, there are specific nutrients that improve cell communication.

7 Telomeres become shorter

At the end of your chromosomes are tiny 'caps' of DNA that have been likened to the caps at the end of shoelaces. Every time a cell divides these caps – or telomeres – become shorter. If these telomeres shorten too much or become damaged, the cell dies or becomes senescent.

You will learn how particular foods and nutrients can help maintain the length and health of your telomeres.

8 The body fails to sense nutritional intake properly

This not only leads to people becoming overweight, but to a blunted reaction to key hormones like insulin – leading to diabetes and many other diseases. And the body ratio of fat to muscle increases. However, it is possible to significantly improve your cells' ability to sense when nutrient levels are inadequate or excessive. That's key, not just to ageing, but to maintaining a healthy weight.

9 Proteins accumulate errors

We think of proteins as part of what we eat. In fact, you *make* thousands of different types of proteins and these proteins do most of the work in your cells. They transmit signals, move oxygen around the body, create structures like collagen, create immune antibodies, and read the genetic code stored in DNA.

But if proteins become misshapen, they cannot function properly. Organs malfunction, bones weaken, immune function declines. We will find out how to reduce the level of protein error, which otherwise will increase over time.

To the original nine *Hallmarks of Aging*, I have added a tenth. One that many researchers are working on:

10 The microbiome becomes unbalanced

Your microbiome is the collection of microbes living in your gut. When the ratio of good to bad bacteria goes out of balance, the results are a poorer metabolism of food, a weakening of the immune system, and the surfacing of many common health problems.

Because the gut and brain are directly linked, the latest research also shows that, if your microbiome becomes unbalanced, mood and brain function can be directly and adversely affected.

However, you can improve your gut health by including more fermented foods and fibre in your diet, with a Mediterranean style diet and sometimes with a course of probiotics.

There are universities and pharmaceutical companies working hard to develop drugs to stop, prevent or slow each one of these ten reasons why we age. The attraction of drugs to researchers is that they are 'silver bullets' – studies identify a problem, and the drugs hit that specific target. They are usually very profitable, which helps pay for the research.

But drugs quite often come with side effects – so we will try to answer a more exciting question:

What if we tackled all the hallmarks of ageing simultaneously?

Not with drugs, but with food and nutrition and some simple-to-add lifestyle changes?

I have included a guide on how you can stay fit with just 30 minutes of activity a day, and one on de-stressing techniques, because mental wellbeing is every bit as important as physical health to ageing well.

Delay Ageing also includes a chapter on how to protect yourself against Alzheimer's, because there is a lot you can do to decrease the risk of this most feared disease.

Everything is connected

You will see, as we unravel the latest science – much only published in 2020 – that all these markers of ageing are interlinked, as are the solutions. Because the causes of ageing are so varied, ways to counteract them must be equally comprehensive.

Is it true that, as one prominent researcher has claimed, there is a single master group of genes that control all the process behind ageing? A group of genes called SIRT genes?

Although we will be looking at SIRT genes, the great majority of researchers believe that ageing is far more complicated than a single set of genes which can, possibly, be manipulated. That was the conclusion of a major 2019 paper in the leading scientific journal *Nature*[2].

But hold fire!

As you read all the ways to delay ageing, you will see many of the same foods and nutrients and solutions mentioned again and again. By all means, make a tentative list but hold back until you get to Chapter 17, where we will summarise an easy-to-follow plan to delay ageing.

You are, literally, amazing

Before we start, pause a moment to reflect on how incredible your body is.

Look at that dot in the middle of the box on the right. You can hardly see it, which is why I added the arrow. Yet it's a little bigger than the very largest cell in a human body. The egg. All other cells are so small that they are invisible to the human eye.

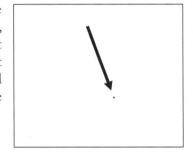

Now look at this illustration of a human cell enlarged hundreds of times.

Into that tiny, tiny dot in the square above are packed the cell's membrane, which includes sensors that communicate with other cells, its mitochondria, which produce energy, and its ribosomes, that produce the thousands of proteins every day that make your body work.

But even that is not what is so incredible. Almost in the middle of the cell, you will see the round nucleus. In that nucleus are packed your chromosomes (the little 'X' shapes), and on those chromosomes are your DNA.

If you stretched out the DNA in a single cell, it would be 2 metres long on average!

Around the outside of that microscopic cell – and inside too – are receptors. These continuously sense and interpret the chemicals in your blood stream like hormones, vitamins, minerals, toxins, pharmaceutical drugs, and trigger an appropriate response. Each sensor does that millions of times a day.

The scale is almost unimaginably small and yet the daily activity level within every one of your cells is almost inconceivably huge.

The average body has an estimated 37 trillion cells – 37 million million. Those cells make tissues and tissues make organs. So, your health and ageing are ultimately controlled at the level of the cell.

About 50 billion cells die and are replaced each day. All those new cells need the best possible nutrition if they are to function well and not to build tissues that age prematurely. That's our focus.

Our objective for healthy delayed ageing

Three quotes help sum up our objective.

The first is from Dr Nir Barzilai, Director of the Institute for Aging Research at the Albert Einstein College of Medicine who says:

"Death is inevitable, but aging is not".

Well, chronological ageing *is* inevitable, but we don't have to suffer the generally accepted biological consequences.

The second is from age researcher Dr Corinna Ross, a biologist at Texas A&M University in San Antonio:

> *"I'm not interested in creating a population that lives to be 150, because that would be a problem for the world we live in.*
>
> *"But if we can keep people out of nursing home care and reduce the number of Alzheimer's and Parkinson's patients, that would be ideal."*

The third is from Dr Brian Kennedy, former president of the Buck Institute for Research on Aging in Novato, California.

> *"We're better at keeping people alive with the various diseases of ageing, but we rarely bring them back to full health …*
>
> *"I think it's going to be much better for the quality of life of the individual, and much better economically, if we can just keep them from getting sick in the first place."*

Prevention is <u>far</u> better than cure.

Unfortunately, that's not the current model of our healthcare systems. Although, of course, doctors, hospitals and pharmaceutical companies want you to get well, the current model is that they mostly only get paid when you are sick.

It's time we changed

We can visualise our objective in two timelines. Without intervention, this is the life pattern the average person can expect:

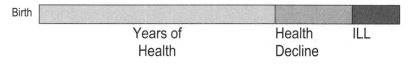

Birth

| Years of Health | Health Decline | ILL |

LIFE PATTERN WITHOUT INTERVENTION

This, realistically, is what our life pattern could be, based on the evidence of the hundreds of studies behind this book.

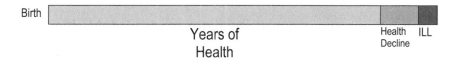

Birth

Years of Health

Health Decline   ILL

WHAT OUR LIFE PATTERN COULD BE

The Big Picture

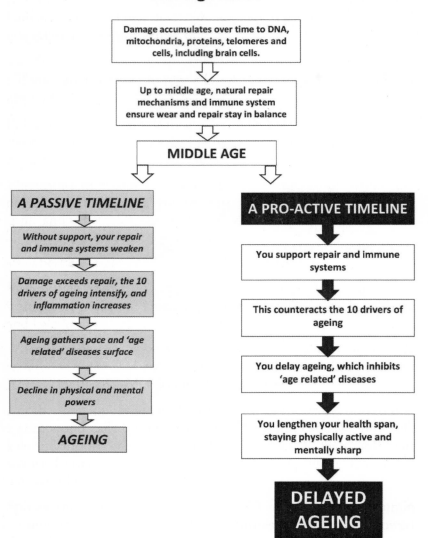

PART I

THE DRIVERS OF AGEING

Here are the 10 markers – drivers – of ageing in visual form. These are the obstacles we have to overcome.

We will investigate each in turn in Part I.

- 1 -

Curse of the zombie cells

Investors in biotech companies are putting billions into a new category of drugs that promise to extend the number of years you can stay healthy – and improve the quality of life in later years.

Dr Richard Miller is a leading gerontologist at the University of Michigan. In an article in New Scientist headed 'The Age of Living Younger', he states:

> *"We have proven you can slow the ageing process using drugs."*

He's right. But Big Pharma's expensive drugs are not the only – or even the best way. Because you can get the almost the same result with nutrition, food and exercise.

Toxic 'zombie' cells need to be cleared away

The trillions of cells in your body are continually turning over. They replicate and ultimately die. An estimated 50 billion die each day.

A normal cell death occurs through a process called apoptosis, and those dead cells are usually cleared away by the immune system – through a vital process called autophagy.

Successful autophagy – *auto* meaning 'self' and *phagy* 'eating' – is critical to health, and it is a subject we will return to in other chapters.

However, occasionally the dead cells or debris from the dead cells are not cleared away. These are called senescent cells – cells that have suffered irreversible damage, but which have not been eliminated from the body. The number of senescent cells increases over time, linked to the weakening of the immune system.

Scientists sometimes refer to senescent cells as junk or zombie cells. Very recent research by Judith Campisi at the Buck Institute for Research on Aging shows that these senescent cells secrete a cocktail of damaging chemicals that poison surrounding tissues and cause chronic (long-term) inflammation. This inflammation is a major driver of ageing, heart disease, arthritis and cancer.

Inflammation

You will be familiar with acute (short term) inflammation. You get a cut or insect bite and your immune system gathers its troops and floods the spot with white blood cells. A symptom is redness and swelling, which shows the white blood cells are in place and working.

That type of inflammation is a necessary, healthy part of repair. After a little while, the inflammation disappears because the immune cells have completed their job.

Chronic inflammation is different. Internal damage to cells and tissues causes the same initial immune response, but the immune response does not entirely stop and remains at a low level, so inflammation persists. It becomes chronic.

We know that this chronic (long-term) inflammation underlies almost all age-related diseases – cardiovascular disease, diabetes, cancer, depression, Alzheimer's disease, and the visible and invisible signs of ageing.

Lynne Cox, a biochemist at Oxford University, summarises the threat from the senescent cell-inflammation connection:

> *"Senescent cells are very bad for you – they destroy the tissues around them."*

Even a small number of uncleared senescent cells triggers problems – just as a rotten strawberry in a punnet can cause the rest to go bad. Indeed, hold that analogy in your mind, as it will have special significance as we unravel this latest science.

The huge cost of senescent cells

The damaging sequence – senescent cells poison surrounding tissues, which causes chronic inflammation, which causes multiple later-life illnesses – is central to ageing. The result is that the average person will live the last 12 to 18 years of their life with some sort of health issue and increasingly on medication.

This is not only distressing for the individual, but a huge cost for society. Some 70% of healthcare costs in the US are linked to later-life degenerative disease. It's similar in the UK where the annual spend on healthcare is £2,892 for every man, women and child and a large part of that is spent on later life.

So, a drug that clears senescent cells and therefore helps extend healthy life span could literally save trillions.

Clearing senescent cells is one of the most important routes to a longer, healthier life. As one gerontology researcher put it, perhaps too bluntly:

> *"We are not trying to fill care homes with geriatrics, we are trying to make sure you live actively and well to enjoy your pension to the full!"*

You can see why bio-tech investors are so excited. But our interest is in doing this without Big Pharma's patented pills.

How to clear away senescent "zombie" cells

If you take senescent cells from an old mouse and transplant them into a young mouse, the youngster ages prematurely, declines physically and cognitively and suffers age-related diseases like atherosclerosis and dementia.

The reverse is true. In 2011, researchers at the Mayo Clinic used a drug to clear senescent cells from aged mice – and it significantly extended the time they were disease-free and enabled them to function as young mice[3].

17

Removing senescent cells *"jump starts tissues' natural repair mechanisms"*, as a Johns Hopkins researcher puts it.

Other studies in 2016 and 2018, including at the UK's Newcastle University, showed that a range of illnesses – from simple frailty to heart disease, osteoporosis, dementia and even osteoarthritis – was halted or even reversed by clearing senescent cells[4].

As a leading researcher put it, we are now seeing:

> *"… how multiple diseases of ageing are being caused by a single underlying cause".*

It is a breakthrough in seeing how healthy lifespan can be extended.

James Pickett is head of research at the Alzheimer's Society, and was not involved in the Mayo Clinic study[5]. Commenting on the finding that clearing away senescent cells also appears to help prevent cognitive decline and dementia, he said:

> *"There hasn't been a new dementia drug in 15 years, so it's exciting to see the results of this promising study."*

The importance of senolytics

So how do we clear away senescent cells? The answer is with senolytics – defined as molecules that seek out, destroy and eliminate senescent cells.

There are drugs that appear to do this – like the cancer drugs dasatinib and navitoclax and possibly the diabetes drug metformin – but none appears to work very well on their own, without being coupled with flavonoids.

Flavonoids are a class of nutrients in fruits and vegetables that are as important to health as the vitamins and minerals that fruits and veggies also provide. They include the pigments in plants – like lutein, lycopene, and beta carotene. Many flavonoids also have anti-cancer properties, as well as being senolytic.

Fisetin – a powerful senolytic flavonoid

The most powerful flavonoid with senolytic activity appears to be a little-known pigment called fisetin – found in strawberries – at a level that is 6 times higher than in the next fruit, which is apples.

A lead researcher at the Institute on the Biology of Aging and Metabolism at the University of Minnesota told Newsweek magazine:

> *"We're looking for drugs that can kill these damaged senescent cells that are very toxic to our bodies and accumulate as we get older. It turns out that fisetin is a natural product that we were able to show very selectively and effectively kills these senescent cells[6]."*

You will note that she didn't suggest you just loaded up with strawberries. Unfortunately, the default reaction from too many researchers is to make a natural product into a drug – because that's patentable and profitable.

Spermidine

As a health writer, I read hundreds of scientific papers a year. But I have never, until recently, read one that announced research data that was "new and thrilling"! This from a journal with the less than heart-racing title: *Biochemical Society Transactions*.

This 2019 article "Spermidine: a physiological autophagy inducer acting as an anti-ageing vitamin in humans?" reviewed recent studies on a natural organic compound called spermidine[7].

Spermidine is found, in order of level, in wheatgerm, soybeans, mushrooms (especially shiitake), blue cheese, aged cheddar cheese, peas, some nuts, and fermented foods including sauerkraut and miso. The compound was first identified in semen – hence the rather unedifying name.

In fact, spermidine – according to one of the original studies published in *Natural Medicine* -- is one of the new longevity and health-span star molecules. Lead author Tobias Eisenberg states:

> *"Oral supplementation of the natural polyamine spermidine extends the lifespan of mice and exerts cardioprotective effects.*[8]*"*

Food consumption studies in humans confirm that low spermidine intake correlates with high heart disease and stroke risk, but this risk declines as the average intake of spermidine increases.

Other research published in *Gerontology* shows that the longevity effect is because spermidine promotes both healthy autophagy and acts as a senolytic[9].

Spermidine is also produced naturally in the microbiome – the gut. Although the human body – male and female – makes spermidine, levels decline with age. So, some age researchers suggest that a probiotic supplementation my help boost levels of polyamines like spermidine and the most promising strain of probiotic boosting polyamines appears to be Bifidobacterium lactis[10].

Exercise as a senolytic

A report in *Diabetes* in 2016 indicated that exercise can help reduce the number of senescent cells, whereas a high fat and high sugar diet increases them[11].

We will be detailing a 37-minute a day exercise programme that will help decrease senescent cells in Chapter 14.

Vit B3 as nicotinamide – another proven senolytic

Nicotinamide is the form of vitamin B3, which, unlike niacin, does not cause facial flushing. Nicotinamide has also been shown to act as a senolytic, helping to clear away senescent cells.

There are already supplements on the market that combine resveratrol, quercetin (a flavonoid found in onions and apples) and nicotinamide as a proposed senolytic combination. But they are jumping the gun. Partly because research on quercetin and resveratrol as senolytics, on their own, is not conclusive. But mainly because senolytics are not the whole story.

All-round cellular spring cleaning

To create a truly comprehensive plan that will clear senescent cells and therefore prolong health, we need to include what are called mTOR inhibitors. Yes, it's a mouthful – but stay with me!

The mTOR pathway

mTOR is a pathway within your cells that regulates the life cycle of that cell – its growth, repair and natural death. The science is rather complicated, but the key is this.

mTOR senses your nutrient intake and when it senses adequate nutrients, the cell will replicate.

When it is suppressed, then rather than replicating, the cell goes into repair mode.

When it is over-stimulated it can promote excess growth and even tumours.

It's a finely balanced cycle – sometimes you'll want to suppress or down-regulate mTOR to ensure your cell repair mechanisms are working properly.

One way to suppress mTOR is calorie restriction, which we know improves cell repair and makes organisms live longer. Which is why some people fast.

But starving yourself is hardly the most attractive way to delay ageing and live healthier for longer. Moreover, as we shall see, there are ways to restrict calories that are easy to tolerate!

Reducing mTOR increases cell repair

The key fact to remember is that to increase cell repair – thereby boosting health and longevity – you need to down-regulate or damp down the mTOR pathway.

Assuming fasting is not your first option, there are drugs that do this – possibly again including metformin. Exercise also helps suppress the mTOR pathway, and certain nutrients and food compounds do it. So why risk the drugs?

Natural mTOR inhibitors

Some of the most powerful, natural, mTOR inhibitors include curcumin, EGCG (an ingredient in green tea), omega-3, olive oil, indoles (a compound in broccoli, sprouts and kale), genistein (a compound in soy with anti-cancer credentials) and vitamin E[12].

Other mTOR inhibitors include grapeseed extract, fisetin (again) and quercetin (a natural pigment and flavonoid found in many fruits and vegetables including onions, apples, red grapes, raspberries, cherries, broccoli and green tea).

You should note that there are factors that can dangerously over-stimulate the mTOR pathway, notably excess protein, especially animal protein. This makes sense – if mTOR senses excess nutrition intake, it may go into cell replication overdrive and cancer is marked by unrestricted cell replication.

Waste disposal for old cells

The final piece of the jigsaw is that down-regulating mTOR also encourages the essential process of autophagy which you want in order to achieve a frequent 'internal spring clean'. Or as one researcher puts it – *"to act like a garbage disposal system for old cells"*.

Intermittent fasting also increases autophagy and you can see why[13]. Fasting creates exactly the challenge of 'mild stress' that brings the mTOR process into play.

Never too old

A key researcher in senescent cells is Professor Xu at the Mayo Clinic in USA. He treated mice, whose equivalent human age was 80, with senolytics and saw *"profound changes"* involving extended lifespan and health-span[14].

So, it's never too late to start the **Delay Ageing** plan.

Fill up on strawberries every week

An encouraging point from the senolytic research is that you apparently don't need to act every day. An occasional, probably weekly, clear-out seems to work very well.

So why not eat a large bowl of strawberries once a week? That's certainly no hardship for most people – and canned or frozen are fine out of season.

Chapter SUMMARY

- Cleaning out senescent cells that are dead, but still present in the body, prevents them causing toxicity which leads to inflammation and thus to many diseases that are related to ageing. That includes osteoarthritis, heart disease, arthritis, some cancers and Alzheimer's.

- We need to encourage autophagy because it clears the body of senescent cells.

- Damping down mTOR activity encourages the process of cell repair, which otherwise becomes weaker over time. It also ensures that 'zombie' cells are cleared away.

- You can liken both processes to a regular spring clean.

- There are natural compounds that do all this – including nicotinamide (vitamin B3), the flavonoids in strawberries, blueberries, turmeric (its key nutrient is curcumin), green tea, grapeseed, green leafy vegetables, onions and garlic. The carotenoids like lutein and lycopene and beta carotene also play an important role.

- Exercise is equally important to clear senescent cells.

- 2 -

Repairing damaged DNA

You have an estimated 37 trillion cells – that's 37,000,000,000,000. About 50 billion die and are replaced each day, and the DNA in millions more is damaged every single day.

This damage is inevitable, since the very act of living and breathing creates DNA damage. That's because the biggest cause of DNA damage is when food is metabolised with oxygen to create energy within your mitochondria. This process causes 'oxidative damage' – caused when oxygen reacts with fats and glucose.

You have seen examples of oxidative damage when oxygen reacts with glucose and an apple goes brown. Or when oxygen reacts with fat and an avocado goes brown, or simply when oxygen causes rust.

Oxidative damage is also called 'free radical damage' and free radical damage to DNA and to cell membranes is a direct cause of ageing and is implicated in the hardening of arteries, wrinkle formation and cancer.

Free Radicals

 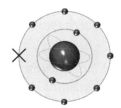

STABLE ATOM FREE RADICAL (missing electron)

When oxygen reacts with fats or glucose, it can cause what is called oxidative damage. The products of that process are free radicals. Free radicals occur when an electron is stripped away from an atom. This is a highly unstable state, because electrons are always paired. Consequently, that free electron urgently 'seeks' another to pair up with. This causes a chain reaction with electrons stealing other electrons and damaging cells and DNA until an antioxidant donates one of its electrons to finally create stability.

DNA damage and cancer

Although damage to DNA in a cell is a primary cause of cancer, it takes time for a damaged cell to become malignant. That's because dozens – even hundreds – of mutations must first accumulate in that cell.

In addition, genes that cause the cell to proliferate uncontrollably (oncogenes) must be switched on. And genes that would otherwise cause the mutated cell to commit suicide (tumour suppressant genes) must be switched off.

For many tumours, like breast or bowel tumours, the whole process can take at least 10 years before they are detectable. For others like prostate cancers, it can be 20, 30 or even more years.

This explains why cancer is linked to age – but also shows that there are steps we can take to interrupt the process and why screening is so important. If most cancers are detected early, they can be treated before they spread – metastasise.

What causes DNA damage?

Other causes of DNA damage include cigarette smoke, pollution, pesticides, UV light, the brown molecules on charred food created by high temperature grilling, excess alcohol, the preservatives in processed meats like salamis and bacon, and red meat whose high iron content triggers oxidation.

Since free radical damage can also be caused by oxygen reacting to glucose, you can reduce your free radical load by cutting down on sugary foods, and on refined carbohydrates which the body metabolises like sugar. And by eating a little less, as eating food raises glucose levels.

Antioxidants in food, and antioxidants created by the body, help to neutralise oxidative damage. Which is why eating foods with a high antioxidant content is part of healthy ageing.

However, although antioxidants are necessary to help protect DNA, they are not sufficient. Indeed, antioxidant supplements like vitamins A, C, E and selenium, on their own, have now been shown to have a limited effect on reducing heart disease or cancer.

Rather than trying to prevent free radical damage, which is difficult, the real key is to support your own DNA repair mechanism.

How to support DNA repair

Your body has powerful, natural DNA repair mechanisms. But, unsupported, they get weaker over time.

Studies have proved that a wide variety of plant foods and nutrients from plant foods help actively support DNA repair and sometimes can even reverse damage. They include:

Plant foods

Foods high in the carotenoids, like beta carotene (in carrots, sweet potatoes, kale and spinach), lutein (in kale, spinach, egg yolks) and lycopene (in cooked tomatoes, grapefruit, red cabbage). A study in the *British Journal of Nutrition* showed that supplementing with mixed carotenoids improved DNA repair mechanisms[15].

Another trial at Lund University, Sweden showed that a combination of nicotinamide (a version of vitamin B3), zinc and carotenoids boosted cells resistance to DNA breaks[16].

Foods high in selenium, like whole grains, brazil and cashew nuts, fish and chicken. Selenium appears to switch on a gene controlled by a protein called P53 which helps repair DNA and suppresses the formation of tumours — as do soy isoflavones and curcumin[17].

Folic acid/folate. A study at the University of Sheffield found that folic acid supplementation reduced DNA damage[18].

A further long-term study of 1,700 people at North Carolina University showed that the combination of selenium with folic acid was colon cancer protective[19].

The Science Officer for Cancer Research UK confirms that folate, a B vitamin, is needed to repair DNA, *"as damage to a cell's DNA can lead to cancer"*.

Folate is found in eggs, leafy greens and whole grains. Folic acid is the form in supplements and is twice as well absorbed as folate from foods.

Flavonoid rich foods. Flavonoids are a large group within the polyphenol class of phytonutrients (plant nutrients). Found in fruits and vegetables, flavonoids are vitally important to health.

Top **fruits** for supporting DNA repair are lemons, raspberries, strawberries, apples, and especially blueberries[20].

Vegetables known to support DNA repair are broccoli, celery, watercress, garlic and cruciferous vegetables like spinach, cabbage and kale[21].

When participants were fed threequarters of a cup of humble watercress a day in a trial at Edinburgh University, they benefited from a significant reduction in DNA damage[22].

Another trial, published in the *British Journal of Cancer*, showed that a compound called indole-3-carbinol found in cabbage, broccoli, kale and another called genistein found in soy (and the supplement soy isoflavones) can increase the level of two proteins that repair damaged DNA[23].

Spices and herbs known to boost DNA repair include curcumin – found in turmeric and therefore in curries. A 2017 study showed that taking curcumin, which has an anti-tumour effect, halved

DNA damage, because it also boosted the activity of an enzyme called catalase which is a powerful neutraliser of free radicals[24].

Ginger, which is a relative of curcumin, also helps boost DNA repair, along with parsley and rosemary.

Within an hour of consumption, green tea boosts the activity of a DNA repair enzyme that goes by the mildly amusing name of OGG1. Another study published in *Carcinogenesis* shows that green tea polyphenols can help cause prostate cancer cells to destroy themselves[25].

Finally, grape seed and grape seed extract have been found to support DNA repair and cause cancer cells to self-destruct[26] [27].

As we progress through this biological detective story, you might wonder why so many of the healing and age delaying compounds are plant based.

Age researcher Frank Madeo puts it well:

> *"There is a million years of coevolution between animals (humans) and plants, which is probably the reason why many of the blockbusters in medical treatment are plant-based substances."*

There is another reason. Plants do not make these flavonoids and polyphenols for our benefit. They have evolved them to protect themselves from damage by the sun and from insects. That's why so many of the most powerfully protective polyphenols are in the colourings and outer skin of the plants. The very part that most processed food throws away!

Sirtuins – the 'master ageing' switch?

David Sinclair is a professor in the Department of Genetics at Harvard Medical School and a leading researcher on ageing.

Dr Sinclair's argument – in his book called *Lifespan* – is that ageing is a disease and that disease is curable. Ageing, he says, is caused by a loss of information as cells are copied over and over throughout life. However, there is a cluster of 7 proteins, called sirtuins, that keep cells stable and help prevent that loss of information.

There are sirtuins that work in the mitochondria and others that work in the cell nucleus. Sirtuins work to maintain cell stability by turning genes on and off – a process that is called epigenetics – which is the subject of Chapter 4.

A wealth of research confirms that sirtuin proteins need a molecule called NAD+. NAD+ (full name oxidised nicotinamide adenine dinucleotide) was first discovered in 1906 and occurs naturally in every cell. It is essential in hundreds of metabolic processes, including creating cellular energy. But levels decline with age[28]. So, scientists have been looking for a way to boost NAD+ levels, as NAD+ improves cells' ability to repair DNA damage.

Increasing NAD+ with nutrients

To create NAD+, your body needs what are called 'precursors', essential nutrients that are needed in the production of NAD+. All the main forms of vitamin B3 can do that, whether niacin (known as nicotinic acid), nicotinamide or nicotinamide riboside.

To increase levels of NAD+, Dr Sinclair personally takes a nicotinamide riboside supplement and adds daily vitamin D, vitamin K, and a low dose aspirin plus a whole gram of resveratrol. Resveratrol is a flavonoid found especially in blueberries and bilberries but also in peanuts, dark chocolate, grapes and red wine.

Finally, he takes 1 gram of metformin – the diabetes drug. Metformin lowers blood sugar levels and increases the body's sensitivity to insulin. However, metformin is a prescription drug, so the average person is unlikely to be able to use it. But, as we shall see, there are other non-drug ways to lower blood sugar levels and improve insulin sensitivity.

Although in animal testing, nicotinamide riboside seems to be a marginally more efficient route to creating NAD+, it is expensive. Moreover, nicotinamide riboside has not been conclusively shown to be better than vitamin B3 as nicotinamide in humans. So, I believe it makes sense to use vitamin B3 as nicotinamide in any supplement that aims to boost energy or repair DNA. Trace food sources of nicotinamide include yeast, milk, fish, nuts, chicken and mushrooms.

There are other ways to boost NAD+ and therefore DNA repair. In times of adequate nutrition, sirtuin proteins mainly support cell reproduction. But when the body is experiencing a challenge – a moderate but non-fatal level of stress – the sirtuins switch from cell reproduction to concentrate on supporting DNA repair[29].

What constitutes moderate stress? Exposure to cold, intermittent (short term) fasting, and short bursts of high intensity exercise. We'll meet these again.

No one mechanism of ageing

Until recently Dr Sinclair's research had focused on sirtuin proteins because they do beneficially turn genes on and off and support DNA repair. However, most age researchers confirm there are multiple mechanisms behind the ageing process – specifically the 10 hallmarks in this book – and boosting sirtuin activity via NAD+ is only one.

There is also a question mark over the effectiveness of resveratrol. Firstly, there are many more polyphenols other than resveratrol that can support DNA repair. Secondly, whilst test tube and mice lab studies on resveratrol do show promise for longevity, these have not yet been confirmed in humans.

Moreover, the 'half-life' of resveratrol in a human diet is very short, meaning it does not remain active in the body for more than 30 to 40 minutes and it is not very bioavailable. (Yes, there is resveratrol in red wine, but sadly you would need to drink hundreds of glasses a day to get a therapeutic dose of resveratrol!).

In 2004, Dr Sinclair formed a company called Sirtris to research resveratrol and the impact of sirtuins on longevity. The idea was to develop compounds based on resveratrol that could be patented. Foods and natural nutrients, thankfully, cannot be patented.

Sirtris was bought by the pharmaceutical giant GlaxoSmithKline in 2008 for $720 million, but it was closed down in late 2013 because the promise of resveratrol-based drugs did not come to fruition.

In 2017, Dr Sinclair widened his focus to include all the hallmarks of ageing we are investigating in this book. In his new company, Life Biosciences, he has partnered with venture capitalists to develop drugs to tackle each of the hallmarks.

Like others, I believe that Dr Sinclair is right to believe that the ageing process can be modified. But foods and lifestyle should be the first focus, not expensive drugs.

Chapter SUMMARY

- DNA is partly protected and is repaired through a wide range of plant foods, vitamins, minerals and polyphenols.

- We have especially noted folic acid, vitamin B3 as nicotinamide, selenium, the carotenoids lutein, lycopene and beta carotene, the flavonoids in green tea, grape seed and curcumin, resveratrol and genistein in soy.

- Plus exercise and what the researchers call intermittent (short term) fasting. For most people, the word 'fasting' is off-putting. So, whilst some small reduction in calorie intake is part of our *Delay Ageing* plan, you will discover an easy and innovative way to achieve that.

- 3 -

Mitochondria – 37 trillion miniature energy factories

A common complaint as people get older is, "I don't have the same level of energy as I used to".

A root cause of fatigue and lack of energy – and one of the 10 key causes of ageing – are dysfunctional mitochondria. Poorly functioning mitochondria lead to neurological, metabolic, muscular and cardiac disorders.

Your mitochondria are the tiny 'power plants' inside almost every cell in the body – except red blood cells. They take in food and nutrients, break them down with glucose in the presence of oxygen and create energy in the form of a compound called ATP.

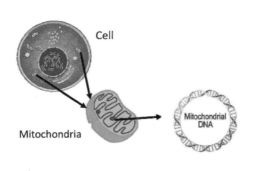

In turn, ATP powers all the millions of biochemical reactions taking place every second in your cells. Within the cell, enzymes accelerate the process of metabolism, creating energy and proteins.

ATP provides the energy for every muscle movement, heartbeat and nerve signal – including brain nerve signals. The fact that the brain uses some 70% of all the ATP produced by the body, explains the strong link between mitochondrial dysfunction and neurodegeneration.

Because the body does not store ATP, the amount of ATP produced by the body is truly enormous. A calculation in the 2016 journal *Science Direct* estimated that the human body creates its own weight of ATP – a day! And in the process, produces some 1,200 watts of electrical energy – enough to power an average dishwasher for 40 minutes.

A by-product of all this activity in mammals is heat – and inevitable free radical damage within the mitochondria, as well as the cell. This free radical damage can lead to mutations, mis-formed proteins and the inability of the cell to produce energy. In addition, the same process can damage (oxidise) fats which enter the blood stream and contribute to atherosclerosis.

If mitochondrial DNA is damaged, the risk of cancer and neurodegenerative disease is increased. A 2014 review article in *Integrated Medicine* confirms:

> *"Mitochondrial dysfunction … is a characteristic of ageing, and, essentially, of all chronic diseases. These diseases include cardiovascular diseases, diabetes and metabolic syndrome.*[30]*"*

The same article goes on to recommend certain supplements that boost mitochondrial repair processes, since preventing mitochondrial damage is difficult. These are discussed below.

Mitochondria have their own DNA

It is believed that mitochondrial DNA – normally written mtDNA – originated millions of years ago when an animal cell literally

mtDNA
[CIRCULAR CHROMOSOME]

engulfed a bacterium. Since this bacterium was efficient at creating energy, it conferred an evolutionary advantage and the animal cell retained it. Hence mtDNA is separate from the DNA in your cell nucleus. mtDNA are circular and you can see the iconic twisted helix inside them.

Interesting, but irrelevant – Mitochondrial Eve

mtDNA hardly changes at all over the generations and is inherited exclusively from the mother. This has enabled genealogists, it is controversially claimed, to trace humans back in time for an estimated 200,000 years, to one hypothetical woman nicknamed 'Mitochondrial Eve'. She may have lived near an oasis in what is modern day Botswana.

Preventing mitochondrial DNA damage with nutrients

In 2017, a research group from the University of Sheffield reported the results of a 5-year study[31] on how mitochondria can protect themselves against DNA damage and repair it. It involves a protein called TOP1 and other researchers have found that this repair mechanism needs a full range of vitamins and minerals and particularly two nutrients called CoQ10 and PQQ.

CoQ10 (co-enzyme Q10) helps protect mitochondria from damage[32] and the lesser known PPQ (pyrroloquinoline quinone) activates genes that help form new mitochondria in ageing cells[33].

In 2017, the Mitochondrial Medicine Society issued a 'Consensus Statement' in which they supported the use, for patients with mitochondrial disease, of: CoQ10, alpha lipoic acid, vitamin B2 (riboflavin), folic acid, and L-carnitine[34].

If cells lack CoQ10 or L-carnitine they cannot produce enough ATP, meaning that muscle fibres tire easily and the heart beats less powerfully – both key signs of ageing.

Co-Enzyme Q10 (CoQ10)

CoQ10 plays a central role in the production of ATP – but levels in the body decline with age. It acts as a powerful protective antioxidant in the membrane of mitochondria. CoQ10 also works synergistically with vitamin E to enhance the antioxidant effect of both nutrients.

A 2014 study found that CoQ10 supplementation *"enhanced mitochondrial activity by increasing levels of the SIRT1 gene"*. As the SIRT1 gene forms the sirtuin proteins that are central in slowing ageing and decreasing age-related disorders, we can see another part in the delaying ageing jigsaw emerge[35].

CoQ10 has also been used to treat and improve heart conditions. Some physicians also recommend CoQ10 supplementation to counteract the adverse effects patients can experience when taking statins[36].

CoQ10 is available in supplements, and in small amounts in peanuts and lentils and in fatty fish like sardines, herrings and mackerel.

PQQ

This powerful antioxidant has been shown to work synergistically with CoQ10 and good sources are parsley, green tea, green peppers and eggs.

Vitamin B3 as Nicotinamide

New research shows that nicotinamide improves mitochondrial quality by causing dysfunctional mitochondria to break down and disperse[37]. This is called mitophagy and is the exact parallel to the important process for clearing damaged cells called autophagy.

Omega-3

Found in fatty fish like salmon, mackerel, herring and sardines, omega-3 helps protect the cell wall of mitochondria. There are plant sources of omega-3, but it is in a form that is less well absorbed called ALA (alpha linolenic acid). Omega-3 helps protect mitochondrial function in the brain and is a nutrient that should be part of any anti-dementia protocol[38].

Alpha lipoic acid

This is involved in breaking down carbohydrates to create energy and is an antioxidant. Alpha lipoic acid is made in small amounts by the body but foods rich in it include leafy greens – broccoli, sprouts, spinach – carrots, flaxseeds, soy beans and rapeseed oil.

In addition to its role in creating energy, alpha lipoic acid is sometimes recommended as a supplement to relieve nerve related symptoms of diabetes. It is known to boost the body's own production of glutathione, a key natural antioxidant, which is important in clearing the body of excess heavy metals. Improved glutathione levels also improve insulin sensitivity.

L-carnitine

The role of L-carnitine is to transport fatty acids from food into the mitochondria to be burned for energy.

Like alpha lipoic acid, L-carnitine helps increases natural levels of glutathione. Although most people produce enough L-carnitine in their own bodies – providing they have enough vitamin C – meat, fish and milk are the main food sources. So, older people on a low or zero meat diet sometimes add an L-carnitine supplement.

The evidence for supplementing with L-carnitine for most ages is not very strong. However, a randomised, controlled clinical trial on 70 centenarians, who were treated with L-carnitine for six months, showed significant reductions in physical and mental fatigue, and an increased capacity for physical and cognitive activity[30].

However, there are reasons why supplementing with L-carnitine for most people might adversely affect the heart, which we explore later. And although the main sources are animal products, you can synthesise L-carnitine from vegetable protein, providing there is plenty of vitamin C in the diet, so most people have adequate levels.

Other nutrients

Other nutrients that are essential for mitochondrial health include:

Vitamin E – found in nuts, wheat germ, seeds like sunflower seeds, spinach, broccoli.

Selenium – found in nuts (especially brazil nuts), seafood, meats.

Zinc – in shellfish, meat, beans, seeds, nuts, eggs and whole grains.

Curcumin – a study in 2016 showed that 4-week supplementation with curcumin was enough to restore mitochondrial function in kidneys and livers, damaged through diabetes[39].

The whole **vitamin B complex** (which includes folic acid) is essential in the conversion of food into energy.

Finally, the active ingredient in **green tea** called epigallocatechin-3-gallate (or **EGCG**) activates a number of proteins including the sirtuin-1 gene which are essential for proper mitochondrial function[40].

Could restoring mitochondrial function improve skin tone and reduce wrinkles? A 2018 study in mice[41] showed that when certain genes were expressed through diet, mitochondrial function was restored, with wrinkles smoothed out and even hair loss reduced.

Chapter SUMMARY

- The full range of vitamins and minerals is needed to support mitochondrial health and repair, both of which are critical to delaying ageing.

- Of particular importance are CoQ10, B3 as nicotinamide, omega-3, curcumin and green tea.

- Older people might respond to additional alpha lipoic acid and possibly L-carnitine.

- Foods to feature are fatty fish, broccoli, sprouts, spinach, nuts, and whole grains.

- 4 -

Directing your genes

In this chapter, you will meet the one-legged cyclists who have proved you really can change the way your genes work.

Your genome is the complete DNA blueprint which contains the unique instructions for building YOU. Or, more accurately, for building the proteins that carry out all the actions that make you function.

It is true that your genome is fixed. But, and it's a big 'but' for healthy ageing, genes can be and are switched on or switched off. They can become either active or inactive. Only a fraction of your 20,000+ genes are active at any one time; they are mostly inert.

This switching activity is called 'gene expression'. Your genetic code remains unaltered, but the way cells read those genes – the way genes are expressed – is changed.

These modifications are triggered by the foods you eat, the activities you engage in, the sleep you get, the environment you live in and the stresses you suffer. Even whether you work nights[42].

Epigenetics – the process of gene expression

Think of the process like this – an explanation provided by Professor Levine of Yale School of Medicine. You have a kitchen full of ingredients and a cookbook. Your genes are the ingredients and epigenetics is the cookbook. The same ingredients with a different cookbook will produce a different outcome.

Genes are nature, epigenetics is nurture. A different environment will produce a different outcome – be it for a child or a gene.

Yet another analogy is from Nessa Carey's book 'Epigenetic Revolution'. Think of human lifespan as a very long movie. The script is like DNA, but if two directors were to take the same script, some scenes would be deleted or modified, and the outcome would be different, for better or worse. In the same way, although your DNA is an unchanged script – you can direct a better outcome.

You are not a prisoner of your genes.

Perhaps the best comparison, though, is with a dimmer switch. You can turn genes down and up and increase their effect, rather like you can adjust your lighting.

But how do genes become active or silenced? This illustration will help.

Inside the Cell

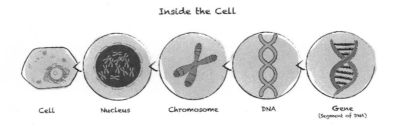

Cell   Nucleus   Chromosome   DNA   Gene
                                                        (Segment of DNA)

Inside the cell is the nucleus, inside the nucleus are chromosomes, chromosomes are made up of DNA and on the DNA are genes.

A gene is activated or suppressed when a 'tag' is put on – or taken off – the gene. This changes the way it functions. '

But how are the tags put on or taken off the gene?

The answer is by what are called methyl donors. Methyl donors are generally nutrients and the process of tagging is called methylation. It is this methylation that can turn genes either on or off.

So, for example, there are tumour suppressant genes you want to be turned on and cancer genes (called oncogenes) you want to be turned off.

There is even a gene called *klotho*, which, when expressed, extends life in mammals. Klotho was a Greek goddess who determined life span by spinning the thread of life.

Certain nutrients are central in this gene expression process. They are called methyl donor foods and include:

Choline found in eggs, fish and meat, flax seeds and lentils.

Methionine also found in eggs, fish and meat but also tofu, cheese, nuts, beans and whole grains like quinoa.

Folic acid found in leafy green vegetables, citrus fruits and eggs.

Betaine found in spinach, beets, wheatgerm and brown rice, and **vitamins B6** and **B12.**

Betaine from food or a supplement is an especially good methyl donor. Folic acid – the supplement form of folate – works with betaine to produce a strong effect on positive gene expression.

It may be better to get methionine more from plant than animal protein as some recent research suggests that this reduces cancer risk and extends longevity – at least in animal experiments.

Methyl donor foods are also involved in the production of several brain chemicals (eg. dopamine and adrenaline) that are involved in alertness, concentration and mood.

Although the science behind gene expression or epigenetics is comparatively new, it is thought that impaired methylation increases with age. Indeed, ageing researchers are now using methylation markers to assess your biological as opposed to chronological age.

Note that methylation and gene switching can also take place on proteins called histones, which wrap around DNA, but the process is complicated and the conclusions for your health are no different.

We do know that impaired methylation can lead to accelerated ageing, depression, fatigue, histamine intolerance, increased risk of cancer, and hormone imbalance. It also leads to birth defects, which is why supplementation with folic acid, a key methyl donor, is recommended during pregnancy.

Eating to direct your genes

All mammals – including you – have a gene called *agouti*.

In a famous experiment[43], when a mouse's agouti gene is completely unmethylated, its coat is yellow, it is obese and prone to diabetes and cancer. When the agouti gene is methylated, however, as it is in normal mice, the mouse has a sleek brown coloured coat and the mouse has a low disease risk.

Yet, fat yellow mice and skinny brown mice are genetically identical. The fat yellow mice are different because they have a harmful epigenetic mutation.

The changes in the expression of our DNA that will favour either health or disease are, to significant degree, under our direct control.

Nutrient deficiency is one of the primary causes of impaired methylation. We have seen that B12, betaine and folic acid are direct methyl donors, but other nutrients play an indirect role. They include zinc, magnesium, potassium, B2, B6, B3 as nicotinamide and sulphur.

Sulphur-rich foods support methylation and include garlic, onions, leeks, eggs and sulforaphane-rich vegetables like broccoli, sprouts, kale, watercress and cabbage. Sulphur is important as it is also a key component of glutathione.

Glutathione – the 'master antioxidant'.

Glutathione has been called the 'master antioxidant', as it recycles other antioxidants like CoQ10 and vitamins C and E. It is a particularly powerful antioxidant, able to prevent free radical damage to cells directly and indirectly.

Glutathione also improves insulin resistance, supports immune function, and supports the detoxification and elimination of heavy metals by the kidneys and liver. Glutathione assists in the important process of ensuring cells die off when they lose function – the process known as apoptosis.

Glutathione is made by the body, but levels normally decline with age. Other nutrients, in addition to sulphur, that are important in boosting glutathione production include selenium, vitamins D3, E, and vitamin C.

A study published in the *American Journal of Clinical Nutrition* showed that taking a vitamin C supplement at 500 mg a day boosted glutathione levels by 47%. Vitamin C elevates red blood cell glutathione in healthy adults[44] and numerous studies have shown that curcumin extract (the concentrated form of turmeric) also significantly increases glutathione levels and thereby reduces the risk of cancer[45 46].

Taking glutathione directly as a supplement, however, has little effect as it breaks down rapidly when taken orally.

Epigenetics and cancer

Dean Ornish is a clinical professor of medicine at the University of California, San Francisco and a pioneer in the effect of nutrition on gene expression and therefore health.

As long ago as 2005, he fed 31 men who suffered from low-grade prostate cancer a plant-based, low fat diet. A control group ate their normal diet.

The study group men were encouraged to walk, meditate to mind-calm, and meet in group sessions. At the end of only three months, some 453 genes, especially ones that controlled for tumour growth, were less active. Overall, blood tests for prostate cancer activity improved and tumours shrank. After 5 years, a check revealed that their telomeres (another longevity marker) were longer than the control group[47].

Although this study was small and involved only men and prostate cancer, there are several indications that the same protocol can help treat breast cancer.

Epigenetics and heart disease

In 2017, the *Journal of Geriatric Cardiology* pulled together decades of research on heart disease. It was highly critical of the usual medical approach to treating heart disease – noting that it was almost entirely focused on treatment and hardly at all on prevention. It is worth quoting from the report:

> *"The side effects of our plethora of cardiovascular drugs include the risk of diabetes, neuromuscular pain, brain fog, liver injury, chronic cough, fatigue, haemorrhage, and erectile dysfunction. Surgical interventions are fatal for tens of thousands of patients annually. And stents carry a 1% risk of fatality.*
>
> *"Patients continue to consume the very foods that are destroying them – the Western diet – which consists of added oils, dairy, meat, fowl, and sugary foods and drinks. This disastrous illness (Coronary Artery Disease) need never happen if we follow the lessons of plant-based cultures where CAD is virtually non-existent.[48] "*

But what is it about a largely plant-based diet that is so cardio-protective? That same meta-study included a trial on 63 individuals with heart disease who followed the Ornish, largely plant based, programme. It compared them to a group of 63 people who did not follow any particular dietary programme.

While the control group experienced no improvement in health, after 12 weeks the Ornish group had lost weight and their blood pressure fell by about 10%. After a year, the activity of 143 genes that encouraged inflammation and injury to blood vessels was significantly reduced. The control group showed no improvements.

We already knew that fruits and vegetables deliver a protection effect against heart diseases as well as cancer and diabetes. Now we know one of the reasons why. They contain compounds (flavonoids and polyphenols) and vitamins and minerals that switch on protective genes and switch off genes that lead to harm.

These flavonoids and polyphenols not only extend health, they have been shown to reduce depression via epigenetic change. In one 2018 study, grapeseed extract polyphenols altered gene activity to create better resilience against depression[49].

We have already noted the fact that blueberries can reduce DNA damage – it appears that they do that by inhibiting a gene called MTHFR[50].

Finally, the damaging effect of air pollution is frequently in the news. Pollution has been shown to alter tags on DNA that increase the risk of neurodegenerative disease.

Reports that supplementing with B vitamins could help protect against these risks were initially met with scepticism – but now we know that the B vitamins exert a positive epigenetic effect, which could be the mechanism for this protection.

Gene expression in the one-legged cyclists

We all know that exercise is good for you, But why? Scientists at the Karolinska Institute in Stockholm have part of the answer[51].

They recruited 23 young and healthy men and women, took them to a lab and asked the test subjects to cycle using only one leg. The cyclists did not use the other leg.

Both legs would therefore experience changes in any methylation patterns from diet or environment; but only the leg that was pedalling would show any epigenetic changes due to the exercise. The volunteers pedalled one-legged for 45 minutes, four times per week for 12 weeks.

Unsurprisingly, the volunteers' exercised leg was more powerful at the end than the other leg, showing that the exercise had resulted in a physical improvement. But it was the changes within the DNA of the exercised legs' muscles that was the surprise. New methylation patterns had occurred in over 5,000 sites on the genome of the muscle cells on the exercised leg.

The genes that had been switched on were those that enhanced energy metabolism, better insulin response and reduced inflammation. In other words, the changed gene expression directly affected health of their bodies. None of these changes occurred in the unexercised leg.

How your gut affects your genes

There are microscopic bacteria living inside you that can directly affect the activity of your genes.

The make-up of the bacteria in your gut is called your microbiome. It is possibly the hottest topic now in nutrition, because the ratio of 'good' to 'bad' bacteria and the diversity of those microbes, directly affects your health.

Not only are several vitamins made in the gut (like vitamin K, B12 and folate), but recently we have found that the composition of your microbiome also causes methylation to take place and either beneficial or detrimental genes to be expressed[52].

Building a healthy gut microbiota can help bring methylation back in balance. This can reduce the risk of colorectal cancer, and even, in the early years, protect against autoimmune disease developing[53] [54].

"Research is showing that the bacterial microbiota of the gut can place chemical tags on our DNA and influence gene expression, potentially impacting our health and many aspects of our lives." Molecular Cell Journal 2016

Foods that help create gut health and positive gene expression include fermented cabbage like sauerkraut, kefir, miso, and some yogurts.

Friendly bacteria – probiotics – are alive and need their own food sources, which are called prebiotics. Good prebiotic foods include the allium family – garlic, leeks, shallots and onions – together with apples and bananas. Oats, whole grains, and flax seeds (and seaweed) are also good sources of prebiotic fibre.

Eating these prebiotic foods is known to encourage the production in your gut of a compound called butyrate – a so-called 'short chain fatty acid'.

Butyrate is very important for your health. It helps protect the integrity of the gut lining – thus preventing 'leaky gut' syndrome. This is a condition where breaches in the gut lining can allow toxins to leak into the bloodstream, triggering inflammation.

Butyrate reduces this intestinal inflammation that can otherwise lead to IBS and colitis. Butyrate can additionally switch off genes that promote inflammation and switch off other genes that would lead to over-active immune response – which in turn is linked to potentially dangerous allergic reactions and some auto-immune diseases.

Gut flora – the microbiome – and the brain's sleep mechanisms are linked via the two-way gut-brain axis. Very recently, researchers at the Sleep Performance Research Center at Washington State

University have found that increasing butyrate levels significantly improved sleep patterns[55].

Since butyrate is produced in response to dietary fibre, it suggests that one way to improve sleep could be to increase your intake of prebiotic fibre from foods like garlic, leeks, shallots and onions, together with apples and bananas.

There is also increasingly strong evidence that a good quality multi-strain probiotic supplement can deliver many health benefits – including via gene expression. That's a topic for later.

Chapter SUMMARY

- You can directly affect gene expression through sulphur-rich foods that provide methyl donors – like choline and methionine, plus the B vitamins including folic acid and betaine.

- Sulphur-rich foods include garlic, onions, leeks, eggs

- Sulforaphane also positively affects gene expression and is found in vegetables like broccoli, sprouts, kale, watercress and cabbage.

- Your diet should also include prebiotic fibre from foods like garlic, leeks, shallots and onions, together with apples and bananas. Add fermented foods like sauerkraut, kefir, miso, and some yogurts.

- Exercise also creates positive gene expression.

- 5 -

Keeping your telomeres long

In 2009, Elizabeth Blackburn and colleagues Carol Greider and Jack Szostak won the Nobel Prize for Medicine for discovering how chromosomes are protected by telomeres.

Chromosomes carry your genetic material and at the tip of each chromosome is a telomere. Telomeres are repetitive DNA sequences, which protect the chromosome and important genes from being deleted. Telomeres are often likened to the plastic tip on a shoelace.

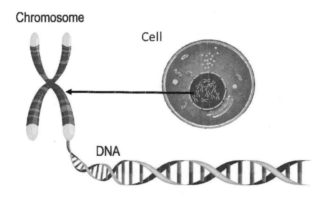

When telomeres get damaged, chromosomes become frayed and can no longer function properly. Cells then cannot renew themselves and they malfunction – they become senescent.

As we have already seen, that means tissues degrade, toxins are released and illnesses like heart disease and cancer surface, your immune system weakens – and ageing is accelerated.

Telomeres become shorter every time our cells replicate, and so they naturally shorten as we age. This continuous reduction of telomere length functions as a "molecular clock" that counts down to the end of cell growth. Just how much they shorten is illustrated by this graph.

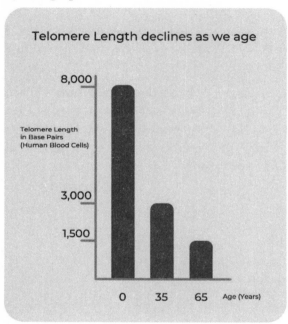

It's based on several sources including a presentation by Lotta Granholm at the Center of Aging at the Medical University of South Carolina[56].

Geneticists at the University of Utah, led by Richard Cawthorn, confirm that shorter telomeres are associated with shorter lives. They found that in the over 60s, those with shorter telomeres were three times more likely to die from heart disease and eight times more likely to die from infectious disease.

Telomere shortening can be reduced

The shortening of telomeres is, unsurprisingly, speeded up by poor diet, a lack of exercise and sleep, by smoking, obesity and stress. However, Dr Blackburn and her colleagues showed that the process can be slowed down and even reversed. The result, says Dr Blackburn, is to keep people healthier for longer and significantly delay diseases of ageing.

To test this, Dr Blackburn put men with low-risk prostate cancer on a low-fat, plant-based diet, coupled with moderate daily exercise, stress management and social support. The lifestyle change resulted in elongated telomeres – while men in a control group had reduction in telomere length. The improvement showed up in a few weeks but was still evident after five years.

Dr Blackburn acknowledges that the shortening of telomeres is only one factor in what needs to be a multi-faceted programme to delay ageing. But she confirms that we have a significant level of control over them all – and the changes do not have to be extreme.

For example, people who do moderate aerobic exercise – about three times a week for 45 minutes – have telomeres as long as marathon runners. Another 2019 study showed the more different kinds of exercise people did, the better their telomeres retained their length[57].

In her book 'The Telomere Effect: A Revolutionary Approach to Living Younger, Healthier, Longer', Dr Blackburn emphasises the role of nutrition. She recommends a mostly plant-based approach that includes flavonoid-rich fresh vegetables, fruit, whole grains, nuts and legumes and high antioxidant foods such as certain seaweeds and green tea.

Since Dr Blackburn's original studies, many other papers have been published on telomeres. Taken together, they indicate that certain nutrients have a special impact to help keep telomeres long:

Omega-3 – found in oily fish and flaxseeds.

Vitamin D3. A study published in 2017 in the *Journal of Nutrition* showed that those with low vitamin D had shorter telomeres than those with adequate vitamin D. It is difficult in the northern hemisphere winter to get adequate, let alone optimum, levels of vitamin D, so a vitamin D3 supplement is advised.

Folate, folic acid and betaine. Many studies confirm it is critically important to reduce the level of homocysteine, an amino acid, in the blood. High homocysteine levels cause inflammation, damage artery walls and are strongly associated with heart disease and Alzheimer's. Folic acid and betaine reduce homocysteine levels.

Vitamin B12. Sub-optimum levels of B12 and folic acid are not only linked to poor DNA repair and methylation, but to shorter telomeres – as are inadequate levels of beta carotene and vitamin E.

Flavonoids and carotenoids. These are found, of course, in fruits and vegetables, so focus on citrus, berries, apples, plums, carrots, green leafy vegetables, tomatoes, nuts, seeds and whole grains. Studies show flavonoids help maintain telomere length[58].

Olive oil. A study published in 2014 in the journal *BMC Medicine* looked at more than 7,200 women over age 55 who were at a high risk of heart disease. Those who consumed the most olive oil, via a Mediterranean-type diet, had as much as a 35 percent per cent lower risk of cardiovascular disease and a 48% reduced chance of mortality over a 10 year period. These improvements were associated with longer telomeres[59][60].

Oats. The soluble and insoluble fibres in oats, whole grains and vegetables help lower blood sugar levels and prevent insulin resistance, which has been shown to shorten telomeres.

Nuts and seeds. A study on 5,500 men and women showed a direct correlation between increased intake of nuts and seeds and increased telomere length[61][62].

So far, then, the story on telomeres seems just a repeat of the advice we already know – eat plants and move more.

But the telomere research has thrown up a dark side of the internet.

Telomerase – a double-edged sword

As a writer, I often receive unsolicited emails about 'breakthroughs'. They typically include headlines like: 'The shoelace secret that promises huge life extension'. Or 'The hand-sized creature that lives 500 years'. Or 'The immortal jellyfish'. (All real headlines!)

The emails, and dozens like them, then try to link me to websites that talk about the need to boost levels of a 'miracle compound' called telomerase. Now it is true that telomerase is the enzyme that is responsible for the maintenance of telomere length. (An enzyme is a molecule that speeds up the rate of chemical reactions that take place inside a cell.)

It is also true that telomerase ensures that telomeres do not shorten prematurely. So much so that the prestigious scientific journal *Nature* is on record stating that: *"Telomerase reverses ageing"*[63].

Telomerase offsets cellular ageing by lengthening the telomeres, adding back lost DNA to put time back onto the molecular clock countdown, which effectively extends the lifespan of the cell. And it does this in adult stem cells. So, the more telomerase the better?

Not so fast.

Although telomerase has undoubted benefits – because it lengthens telomeres – it can also have a harmful side. Just as youthful stem cells use telomerase to offset telomere length loss, cancer cells also employ telomerase to maintain their rapid and destructive growth.

So, trying to boost telomerase levels with a specialised telomere supplement – of course they do now exist – could be risky, as Elizabeth Blackburn herself cautions against untested strategies:

> *"... whereas things like exercise, stress reduction and good diet have never been shown to increase cancer risk, and indeed studies show they decrease those risks".*

Chapter SUMMARY

- Are shorter telomeres a cause of ageing – or just a *symptom* of ageing, which is the result of the other 8 hallmarks of ageing that we started with?

  While the chart at the start of the chapter clearly shows the link between shorter telomeres and age, it is possible that they are more of a marker than a direct cause.

- Other factors – for example chronic (long term) inflammation, DNA damage, adverse gene activation through sub-optimal nutrition, mitochondrial damage and uncleared senescent cells – are probably more important.

- Nevertheless, certain nutrients are specifically associated with maintaining telomere length – they include omega-3, vitamin D, folic acid, betaine, carotenoids, a range of polyphenols, olive oil and oat fibre.

- And, again, exercise!

- 6 -

Reducing the loss of your stem cells

Children recover quickly from injuries or illness, partly because their young stem cells can replace damaged cells and tissues rapidly and efficiently.

Stem cells are 'undifferentiated' cells that can turn into specific cells as the body needs them. In some parts of the body, such as the gut and bone marrow, stem cells regularly divide to produce new body tissues for maintenance and repair. They are therefore a vital part of tissue regeneration and wound healing.

Because stem cells help replace damaged tissue or worn-out cells, their number and activity levels are a major factor in deciding the rate of ageing.

As we get older, however, the number of adult stem cells declines, which impairs their ability to regenerate tissue. Their function is also compromised – which contributes to oxidative stress (free radical damage), DNA damage and inflammation. Moreover, when stem cells divide, their telomeres shorten and cells stop dividing and die.

Short telomeres increase the likelihood of cells becoming senescent and producing toxic molecules that lead to inflammation, which is a major risk factor for age-related diseases. Not for the first time, we are seeing how everything is connected.

To increase the length of time you stay healthy, you need to increase the number and quality of stem cells and to activate the regenerative signals in them.

The drug companies are, naturally, looking to high-tech and patentable solutions like stem cell therapy – where healthy stem cells are transplanted from a matching donor to replace damaged stem cells.

However, there are two, possibly three, ways to restore stem cell function naturally. *Stem Cell Research and Therapy* was able to state in 2018 that:

> *"With current knowledge of stem cells, it is feasible to design and test interventions that delay ageing and improve both health and lifespan. Stem cells in conjugation with anti-ageing genes can neutralize most of the devastating signalling effects which are known to cause premature ageing."*

Can nutrition improve stem cell health?

Certain foods and nutrients may either diminish the impact of stem cell decline or even restore them.

A study in *Stem Cells Developments*, conducted at the Center for Aging and Brain Repair at the South Florida College of Medicine, indicated that a nutritional combination was able to increase the number of stem cells generally, and also increase neuron stem cells, which could improve brain function[64].

The combination tested was vitamin D3 together with polyphenols from blueberries, bilberries and green tea, plus L-carnitine.

However, this was a small study and funded by the company that makes the combination. Whilst that doesn't necessarily invalidate the study, there appears no other subsequent corroborating research. Nevertheless, the nutrients they used feature in several other ways to extend healthy ageing, so they are worth including in a final plan.

Fortunately, there are two other ways to increase stem cells that have convincing evidence for them.

Intermittent 'fasting'

Researchers lead by Valter Longo, Professor of Gerontology from the University of California, have found that fasting for two to four days at a time, every six months, stimulates stem cells to convert from a normal dormant state to start regenerating.

Damaged and senescent cells were destroyed, and new cells created from their components, effectively revitalising the immune system. The researchers also found that levels of a growth-factor hormone called IGF-1 were reduced.

That's important because IGF-1 is a compound that can otherwise promote the formation and growth of tumours, as well as the negative effects of ageing. Reduced levels of IGF-1 were a key factor in triggering the anti-ageing Klotho gene we encountered in a previous chapter[65] [66].

Dr Longo commented that:

> *"We could not predict that prolonged fasting would have such a remarkable effect in promoting stem-cell-based regeneration. If you start with a system heavily damaged by chemotherapy or ageing, fasting cycles can generate, literally, a new immune system."*

But prolonged fasting is hardly a popular plan and not recommended for senior people! So, in a further study, 100 healthy subjects undertook a diet that 'mimicked' fasting by using a plant-based diet, low in protein and simple carbohydrates and totalling 900 calories daily. The participants followed that diet for five consecutive days each month for 12 weeks.

Compared with the control group, the participants using the "modified fast" had a reduction in body weight, in abdominal fat, blood pressure, inflammation, and in the tumour marker IGF-1. Significantly, they achieved a significant increase in stem cells circulating in the bloodstream.

Four months after finishing the three cycles of the fast mimicking diet, continued benefits included the fact that 60 percent of the weight loss was maintained.

Dr Longo explains:

"Fasting forces the body to use its store of fats and glucose. It also breaks down a lot of white blood cells, forcing the body to regenerate new immune system cells".

Later, we give details of a simple, realistic plan to achieve the same reduction in calories as intermittent fasting. One that is easy to build into any lifestyle.

Exercise – the miracle 'drug'

We all know that exercise is good for you – but are you motivated to act on that knowledge? Take 3 minutes to read this section and then decide.

A paper in 2006 in the journal *Stem Cells* showed that exercise increases the number of adult stem cells in aged mice[67].

A further 2012 research paper showed that exercise increases brain stem cells and concluded:

"Regular exercise should be promoted, not only for disease prevention, but to maintain a high reserve of adult stem cells"[68].

An excellent 2017 meta-analysis conducted by the Centre of Health Technologies, University of Pavia, Italy, showed that physical exercise activates both cardiac and muscle stem cells[69]. It helps answer the question – what type of exercise activates and increases stem cells?

The ideal exercise combines moderately intense aerobic exercise (eg. cycling or fast walking) and resistance (strength) training. It is this combination that we have developed in the 37-minutes-a-day activity element of our healthy-ageing programme.

The Health Technologies results also confirmed that exercise creates positive epigenetic DNA changes. This helps explain why people who exercise regularly have a lower risk of cancer and metabolic problems like diabetes – and indeed depression.

Finally, they also confirmed that regular physical activity increases the level of nitric oxide in the vascular system – ie. in blood vessels. This in turn keeps arteries flexible and increases blood flow – which explains why people who maintain physical activity have lower heart disease risk, lower dementia risk and maintained sexual health.

A further study recently concluded at the University of Brighton in the UK showed that exercise significantly improved the regenerative capacity of a type of stem cell – called mesenchymal cells – improving their growth, proliferation and ability to differentiate[70].

This is significant because mesenchymal cells can differentiate to become either bone cells, cartilage cells, muscle cells or adipose (fat) cells.

A study at McMaster University in Canada showed that when mice ran on a treadmill for less than an hour three times a week, it triggered mesenchymal cells to become bone and muscle rather than fat[71].

So, exercise not only creates epigenetic changes in DNA, it is a proven way to maintain a healthy level of stem cells. We are beginning now to see why exercise reduces the risks of cardiovascular disease, diabetes, colon cancer, breast cancer and depression. Of course, it is also a key to weight control.

If a pharmaceutical company had brought out a drug that did all this, it would be a billion-dollar seller!

Blue Zones

A famous study initiated in 2008 by *National Geographic* and carried out by a team lead by Dan Buettner, identified what they labelled 'Blue Zones'. Blue Zones are areas in the world where a significant proportion of the population lives to the age of 100 in good health[72].

The seven Blue Zone areas include the Greek Island of Icaria, the Japanese island of Okinawa, a coastal zone in Costa Rica, the Barbagia region of Sardinia and perhaps unexpectedly, Loma Linda, near Los Angeles, California.

Their common denominators are their diets – a largely plant-based diet, fish, limited occasions for meat and almost no processed food. They all tend to eat their last meal early in the evening and consciously stop eating when they reach 80% full. A practice that Okinawans call **hara hachi bu**.

The other commonalities are that their lifestyle means physical activity is built in. They all have a strong sense of belonging to a community, and a sense of purpose – which the Okinawans call *ikigai* and the Costa Ricans call *plan de vida* – "why I wake up in the morning". They all have a time each day to de-stress.

The apparent odd one out – Loma Linda – is explained by the fact that it is a Seventh-Day Adventist enclave, whose lifestyle is very similar to the others. Except that the others drink moderate amounts of alcohol – mostly wine.

Blue Zone centenarians tend to be sick for only a few months at the end of their life and many simply do not wake up one morning. That seems an ideal way to 'shuffle off this mortal coil'.

Avoiding sarcopenia

Older people commonly lose strength, become prone to falls and resistance to illness and there is an increase in their fat to muscle ratio.

This is medically known as sarcopenia. Sarcopenia is a condition characterized by loss of muscle mass and strength – frailty. It is correlated with physical disability, frequent fatigue and a poor quality of life. Risk for sarcopenia rises with age and low levels of physical activity.

Sarcopenia remains an important clinical problem that impacts millions of older adults. Causes of this condition include a decline in hormones, increased inflammation, reduced activity, and inadequate nutrition.

There are a lot of conditions correlated with sarcopenia like obesity, diabetes and reduced intake of vitamin D. It has been proposed that excess energy intake (aka over-eating), physical inactivity, low grade inflammation, insulin resistance and changes in hormonal homeostasis may result in the development of sarcopenic obesity.

Chapter SUMMARY

- Your supply of stem cells is limited. You can help protect them by ensuring an optimum intake of vitamin D3, and polyphenol rich berry foods and green tea.

- Exercise and periods of lower calorie intake also play a very important role in maintaining the health and supply of stem cells.

- 7 -

Your four longevity 'fuel gauges'

If your body loses the ability to accurately sense the level and type of nutrition you are eating, it is like having a faulty fuel gauge.

Signals to your body and brain go awry, your body takes in too many calories or too little nutrition, metabolic errors occur, diseases like Alzheimer's and diabetes develop, and ageing is accelerated.

A key example of poor nutrient sensing is glucose intolerance, when the body loses its sensitivity to glucose and ultimately to insulin, which can lead to diabetes.

But how does your body carry out its millions of processes a day and match energy expenditure to energy intake, when the amount, variety and quality of your food and nutrition varies so much from day to day?

Four fuel gauges or nutrient sensing pathways

That's the role of four 'fuel gauges' or, as ageing researchers call them, 'nutrient sensing pathways'. They constantly adjust the way you metabolise your nutritional intake to create a stable bodily environment. You can also think of them rather like the sensors on a self-driving car which prompt constant adjustments to keep the car moving ahead safely.

These four nutrient sensors try to ensure you take in just the right amount of macro-nutrients (fats, proteins and carbohydrates) and just the right amount of micro-nutrients (vitamins, minerals and plant nutrients like flavonoids and polyphenols).

When they work properly, you remain in good health, you retain your ability to maintain a healthy weight – and you age slowly.

But things can go wrong. We have already seen that metabolism (the burning of fats and glucose to create energy) inevitably damages cells via the creation of free radicals and mitochondrial dysfunction. This damage also impairs the nutrient-sensors – and the body begins to break down at the cellular level and starts to age. The scientists behind the nine hallmarks of ageing call this process 'Deregulated Nutrient Sensing.'

Why nutrient sensitivity is so important

In our evolution, the body's ability to sense changes in nutrient availability was critical. In times of abundance the body was able to grow and store energy. In times of scarcity it switched to a conservation mode – which increased resilience to disease.

So, you want your body to experience periods of challenge to trigger that conservation mode, promoting healthy ageing.

1. The insulin sensing pathway

The most familiar example of poor nutrient sensing is the development of glucose intolerance. Glucose intolerance is a term for a metabolic condition which result in high blood glucose – blood sugar – levels. The result is pre-diabetes, or when more developed, type 2 diabetes.

Excess intake of sugars and simple, refined carbohydrates – like white bread, cakes and potatoes which are all metabolised like sugar – increases glucose in the bloodstream. This prompts the body to create insulin to reduce the level of glucose.

If this happens too often, the body becomes increasingly insensitive to insulin, therefore blood sugar levels stay dangerously high, which leads to type 2 diabetes and other illnesses. A majority of adults in the West are probably now insulin-resistant.

High glucose levels also create inflammation in cells and tissues, which leads to accelerated ageing, and they repress 'good bacteria' in the gut, thus increasing the ratio of bad or toxic bacteria to beneficial microbes. Signs of glucose intolerance include thirst, bloating, tiredness and depression.

AGEs are (very) ageing

When you have too many sugar molecules in your system, they bind onto the fats and proteins in cells, a process known as glycation. You might visualise them as 'sugar coated'.

This forms what are called Advanced Glycation End products (appropriately shortened to AGEs). AGEs cause protein fibres to become stiff and malformed, and connective tissue becomes cross linked – think glued together or matted.

Externally, this damage causes skin to lose its youthful elasticity and become wrinkled and saggy. The same cross-linking process can produce cataracts in the eye. Internally, cross-linking causes a stiffening and narrowing of arteries which are risk factors for heart attack and stroke.

There is yet another damaging process going on.

Mitochondria subjected to excess glucose are forced to work harder, as they burn the glucose with oxygen to create energy. Remember that this metabolic process creates free radicals which are damaging to the mitochondria themselves. Damage to the mitochondria is why one sign of diabetes is tiredness.

The process doesn't end there. The effects of the excess free radicals (the process is also called oxidation) prompts cell damage and chronic inflammation. In turn, chronic inflammation, as we have frequently seen, leads directly to incapacitating conditions like heart disease, Alzheimer's and diseases of the pancreas and liver, as well as diabetes.

It's a triple whammy – the excess glucose makes cells abnormal (glycated), and creates excess free radicals, which creates inflammation, which is a key cause of ageing. No wonder scientists in ageing research have called the process 'inflamm-ageing'.

The external signs of glycation show up around your late 40s, when accumulated sun damage, environmental oxidative stress, hormonal changes, and the development of AGEs begins to seriously start the process of ageing.

Reduce and slow glycation

While glycation cannot be entirely stopped, it can be reduced and slowed. The obvious priority is to reduce your intake of sugar, and especially high-fructose corn syrup, which some studies have shown increases the rate of glycation 10-fold compared with glucose.

Although the use of high fructose corn syrup is more common in the USA (because of subsidies), it is used in the UK. It can appear on labels as HFCS, as glucose fructose syrup or even isoglucose. It can be found in biscuits, ice creams, desserts, sweets, fizzy drinks, tomato ketchup and even fat free dressings and fat free yogurts.

On the positive side, green tea[73] and grapeseed extract[74] have both been shown to reduce the glycation process, while stimulating the production of healthy collagen.

A hormone called Insulin-like Growth Factor (IGF-1) regulates cell growth and is an important part of glucose sensing. Refined carbohydrates and sugars, which of course raise insulin levels, also raise levels of IGF-1.

Multiple studies show that a reduction in IGF-1 increases healthy lifespan in mice – and discourages the growth of cancer cells[75] [76].

Foods that improve insulin sensitivity, and which reduce levels of IGF-1, include those that are rich in soluble fibre like beans, pulses oats, apples, peas and citrus fruits. Other foods include carrots (rich in beta carotene), cooked tomatoes (rich in lycopene), flaxseeds (a vegetarian source of omega-3), cruciferous vegetables like brussels sprouts, kale and cabbage, and especially polyphenol-rich berry fruits like raspberries, blueberries and blackberries.

Curcumin (extracted from turmeric), ginger, rosemary, onion and garlic are additional powerful foods that help improve insulin sensitivity and the healthy function of the IGF-1 pathway[77].

This is where metformin may come in

Metformin is the most frequently prescribed drug in the world and is given on prescription for type 2 diabetes. Metformin was originally developed from compounds found in the plant called French lilac. (Trade names include *Glucophage* – or 'sugar eater'.)

Metformin lowers the level of sugar in the blood of diabetics, both by reducing the amount of sugar released by the liver and by improving how the body responds to insulin. In short, it improves nutrient sensing.

So – the argument goes – if metformin can lower blood sugar levels it should reduce oxidative stress (excess free radicals), reduce glycation and reduce inflammation – all of which should increase healthy lifespan[78].

There is evidence for the argument. Dr Steven Austad is Scientific Director of the American Federation for Aging Research. He points out that diabetic patients taking metformin tend to have better heart health than diabetics who do not take the drug. They are also less likely to develop age-related cognitive disorders such as Alzheimer's disease.

Supporters of metformin claim the reason is that patients with Alzheimer's disease tend to have reduced insulin sensitivity in the brain. So, increasing sensitivity should reduce the risk.

Moreover, since reduced insulin sensitivity is linked to increased inflammation and inflammation in the brain is linked to

Alzheimer's, metformin may have the power to slow or stop diseases related to age-related cognitive decline. Diabetes patients also appear to have a lower chance of developing cancer if they take metformin.

Scientists believe the drug works in the mitochondria and Professor Austad says metformin makes those mini-power plants run more efficiently, reducing the release of free radicals. He also says that activation of insulin sensitivity mimics the effect of a low-calorie diet, something multiple researchers have noted can extend life span in laboratory animals.

These are the reasons why Professor Nir Barzilai of the Institute for Aging Research at the Albert Einstein College of Medicine announced a trial in 2019 called TAME (Targeting Ageing with MEtformin).

So, we should all be taking metformin? Not necessarily.

First, it is a prescription drug, so unless your doctor is prepared to prescribe it, you cannot easily obtain it.

Second, the primary purpose of metformin is to lower blood sugar/glucose levels by improving the way your body handles insulin – improved insulin sensitivity. But as we have just seen, this can also be achieved through many foods and in a reduction in calories.

Third, while metformin also targets the other three nutrient sensing pathways – which we are about to discuss – its advocates do acknowledge that: *"to date, there is no evidence for such effects in humans."* Hence the TAME trial[79].

Fourth, a study published in 2019 in the journal *Aging Cell* found that metformin seems to reduce the beneficial effect of exercise. That is important, because exercise itself is a proven and non-drug way to reduce insulin resistance.

Fifty-three healthy, but sedentary, non-diabetic men and women were given metformin and matched with a similar panel that merely took a placebo. Both groups were put on a 4-month programme where they exercised on a treadmill or bike for 3 times a week for 45 minutes each session.

Not surprisingly, all the volunteers had better aerobic fitness at the end. But the metformin group had rather less improvement to their insulin sensitivity than the control group. And, unexpectedly, only the non-exercising placebo group had a significant improvement in mitochondrial health. It appeared that metformin had somehow blocked the expected improvement in mitochondrial performance[80].

Finally, there is evidence that long-term use of metformin results in an increase in the risk of vitamin B12 deficiency in about 10% of patients[81].

People on metformin are therefore advised to be regularly checked for their B12 status and, depending on the result, take a B12 supplement. A B12 deficiency can lead to neuropathy – nerve problems – anaemia and difficulties in making DNA.

Berberine – a similar effect to metformin?

Berberine is a yellow, bioactive compound found in several plants in the berberis family including oregon grape, barbary, goldenseal and tree turmeric.

A 2008 study published by the *US Library of Medicine* randomly assigned 36 adults with type 2 diabetes to either metformin or berberine. It showed they both worked equally well in lowering blood sugar, but berberine was slightly better at lowering blood fats and *"inhibiting fat storage"*. The lead scientists on the study concluded that:

> *"[Berberine] may serve as a new drug candidate in the treatment of type 2 diabetes"* [82].

Although 36 subjects constitute a small sample, berberine has been successfully used in China for hundreds of years as a diabetic treatment. A further 2012 study at the Shanghai Diabetes Institute confirmed that berberine was able to decrease glucose and insulin in the blood and strengthen insulin sensitivity[83].

Diabetics should note that berberine shows a strong consistent ability to lower a marker of average blood glucose levels called HbA1c.

The downside of berberine, however, is that it has poor bio-availability, so effective doses are high. Compared to metformin, there is much less long-term safety data.

The reservations over metformin apply also to berberine – an improved diet should produce the reduced glucose levels and improved insulin sensitivity they both seek to achieve.

Neither metformin nor berberine should be taken by pregnant women.

2. The mTOR nutrient sensing pathway

mTOR is an enzyme and the second and key nutrient sensor – one of our four 'fuel gauges'. (An enzyme is a protein that accelerates the effect of other biochemical processes.) mTOR plays a critical role in regulating cell growth, cell proliferation, and life span.

The University of California at Davis has also done considerable research on mTOR and researchers comment:

> *"mTOR is the engine of growth in childhood but the engine of ageing in adulthood".*

You need mTOR when you are young, because that is when you need cells to proliferate and grow. But later in life, cell growth needs to slow down and stabilise. So high levels of mTOR then are inappropriate.

David Sabatini is a professor of biology at the Massachusetts Institute of Technology (MIT) and a world leader in nutrient sensing research. Describing the importance of the mTOR pathway, he said in 2017:

"We now know that one pathway—the mTOR pathway—is the major nutrient-sensitive regulator of growth in animals and plays a central role in physiology, metabolism, the ageing process, and common diseases.

"The mTOR pathway is activated during numerous cellular processes and when it's deregulated (ie. when restrictions are removed) *the results are increased rates of ageing and diseases such as cancer, neurological illness, epilepsy, and diabetes".*[84]

If high levels of mTOR invoke cell growth, you might expect that this would also link to cancer risk – because a hallmark of cancer is rapid and out-of-control cell growth.

Indeed studies do show that, in almost 100% of cases of advanced human prostate cancers, mTOR is over-activated, and present in higher amounts. Similarly, higher levels of mTOR are found in breast cancer tissues, in lung and brain cancer and appear to be associated with worse survival rates.

Simply put, if you reduce mTOR, you may reduce your risk of cancer and increase your chances of living longer.

That is also because mTOR plays a key role in autophagy – the process by which old or damaged cells are absorbed and recycled by the body, before they can spew out toxins that create inflammation.

So how can you lower mTOR to a safe level?

The key point to understand is that mTOR senses the amount of amino acids your body has available and directs how much protein is made in response. When mTOR senses an abundance, it directs the body to build cells – that's the <u>anabolic</u> response.

When it is inhibited, it goes into conservation mode, cells divide less and the components of old or damaged cells are re-used to maintain energy and extend survival. That's the <u>catabolic</u> response and precipitates autophagy.

This is all part of an evolutionary system which protected our ancestors when food and especially protein was scarce. Inhibiting the mTOR pathway allowed them to survive.

So, a situation of mild challenge to cells prompts the activation of longevity genes. Which is why longevity researchers have identified intermittent fasting, temporary exposure to hot or very cold temperatures and exercise to be a key part of healthy ageing. They provide just the right level of mild stress (challenge) and provoke the body to go into conservation and repair mode.

More on mTOR

mTOR stands for <u>m</u>ammalian <u>T</u>arget <u>o</u>f <u>R</u>apamycin and it regulates the development of proteins and cell growth in response to nutrients, energy levels, and stress.

Rapamycin was initially isolated from a soil sample from Easter Island (also known as Rapa Nui). Subsequently, rapamycin was found to possess immune-suppressive properties – and it is used as a drug to prevent rejection after organ transplants. Since rapamycin also inhibits proliferation of mammal cells, it is being studied as an anti-cancer drug.

So, should we, like some longevity researchers, be considering rapamycin as a life extending drug – since it has been used to extend lifespans in yeast, fruit flies, mice and recently marmosets?

The jury is out. Clinical trials have only shown modest results in humans and side effects from rapamycin derived drugs include hypertension, increased cholesterol levels, possible glucose intolerance, fever, and potentially cataracts and nausea.

A lot of current rapamycin research is aimed at finding how to reduce its side effects, which, in fairness, come from people who already have serious health problems and who usually take it in combination with other drugs. But, so far, researchers do not know whether those same problems would happen in healthier people taking the drug on its own. When there are natural alternatives, why take any risk?

Foods to inhibit mTOR

Food can also be an mTOR inhibitor. An article by Dr Rosane Oliveira, Founding Director of U+C Davis Integrative Medicine, who originally described mTOR as *"the engine of ageing"*, explains it well:

> *"One of the drivers of mTOR appears to be the amino acid leucine, which is found in higher amounts in animal foods (eg. dairy, meat, chicken, fish, and eggs).*

> *"Therefore, to lower your leucine intake (and mTOR levels), you need to either restrict your consumption of animal proteins or, better yet, adopt a plant-based diet."* [85]

Other research confirms that eating plants decreases mTOR activation and provides natural mTOR inhibition. Some of the best natural mTOR-inhibiting foods – according to UC Davis – include cruciferous vegetables like broccoli, green tea, soy, turmeric (curcumin), grapes, onions, strawberries, blueberries and mangoes.

It is not coincidental that these foods also feature in many of the Blue Zone diets.

Reducing animal protein intake (rather than reducing calories) may be for most people an easier solution to decreasing mTOR levels.

There are other two other reasons for increasing plant proteins.

When you reduce the consumption of animal protein, you also decrease IGF-1 levels. We have already seen that IGF-1 (Insulin-like Growth Factor) is a hormone that (like mTOR) regulates cell growth. Refined carbohydrates and sugars, which of course raise insulin levels, also raise levels of IGF-1. Studies show that a reduction in IGF-1 discourages the growth of cancer cells.

The steroid hormone DHEA

Your most plentiful steroid hormone is DHEA, which helps produce other hormones including testosterone and oestrogen. DHEA levels are at their highest levels in our 20s but decline as you get older. This has been linked to ageing, heart disease and depression[86].

So, increasing DHEA levels should be part of an overall anti-ageing strategy.

As a result, some people take DHEA as a supplement. But there is no evidence that this is either effective, or even safe[87]. Instead, follow Dr Rosane Oliveira's advice and follow a largely plant-based diet – which she confirms can boost DHEA levels by as much as 20% after as little as 5 days, compared with a meat-eater diet.

Calorie restriction

Regular calorie restriction – the many researchers involved call it CR – is shown to consistently and significantly increase lifespan in several short-lived animals including those research favourites *C. elegans* worms, fruit flies and mice[2].

But being constantly hungry for a few extra years of life is not a trade many of us would happily make.

Fortunately, there are less drastic ways to reduce calorie intake. For some people, intermittent caloric restriction/intermittent fasting (known as IF) is moderately easy to tolerate.

There are several ways to conduct IF. One (we have already seen) is to limit yourself to about 900 calories a day for 5 consecutive days a month.

Another is to restrict eating to just 8 hours a day – say from 10 am to 6pm. You do not eat for the remaining 16 hours – so it is called the 16:8 method.

Yet another is to fast completely for a 24-hour period each week.

A study, led by M A Wijngaarden at Leiden University Medical Centre, found that intermittent fasting induces *"rapid metabolic adaptations"* that repress certain genes and activate nutrient sensing mechanisms.

In turn, this helps aid in cellular repair, reduces free radical (oxidative) damage and inflammation, and may even help prevent cancer[88].

The success of IF in animal studies is strongly linked to the fact that it reduces the level of insulin in the body and improves insulin sensitivity.

For many people who follow an IF protocol, their objective is weight loss – and they should only do it with a doctor's support. For those whose aim is simply to delay ageing, we will propose in Chapter 15 an alternative way to reduce calorie intake that is far easier to accept.

You should note that the CR methods in the human studies involved decreasing energy intake from food, but at the same time maintaining optimum nutrient intake. That means supporting CR with a comprehensive nutritional supplement.

3. The AMPK nutrient sensing pathway

AMPK is another protein and the third nutrient sensor. It is a 'fuel sensor' present in every mammal cell. It detects the level of energy in cells (the number of ATP molecules) and is the master regulator of energy balance.

AMPK senses when nutrient resources may be scarce or whether your body is in a fasting or semi-fasting state. If it is, AMPK increases the uptake of glucose and fatty acids for energy. It simultaneously decreases the energy hungry production of proteins and goes into a conservation mode.

Remember, you want your body to take up glucose so that excess amounts are not left in the bloodstream, and you want to regularly go into a conservation mode – because that creates a situation that prompts cell repair and is important for healthy ageing.

Conversely, when AMPK becomes less sensitive, due to free radical damage, the result is reduced autophagy (worn-out cells are not cleared away properly), more fat is deposited, and inflammation increases.

So unlike mTOR, which you want to inhibit sometimes, you want to activate and increase AMPK.

But how? By far the most effective is exercise[89][90].

Exercise, as we have seen, also improves insulin sensitivity and positive gene expression. In contrast, inactivity is associated with a reduction in AMPK and a corresponding increase in various disorders including type 2 diabetes, coronary heart disease, Alzheimer's disease and cancers of the colon and liver. To be bed-ridden for as little as seven days leads to glucose intolerance and increased insulin resistance.

Overeating inhibits AMPK, whereas calorie restriction, which lowers blood sugar, can increase the activity of AMPK – and is known to increase lifespan in at least short-lived animals. So, a reduction in calories should boost the sensitivity of yet another one of the four key metabolic regulators.

There are foods that indirectly increase AMPK activity – including soluble fibre foods (oats, apples, peas, beans) and polyphenol rich fruits, especially berries. Nutrients that increase AMPK levels include green tea extract, curcumin, omega-3 and genistein from soy.

4. The sirtuin sensing pathway

The fourth nutrient sensor we have already met – sirtuin genes and proteins. A review paper co-authored by a leading age researcher Leonard Guarente confirms:

> *"Sirtuins function to slow ageing and various disorders associated with ageing, including metabolic diseases, cancer, and neurodegenerative conditions."*[91]

Sirtuins detect when energy levels are low and 'call for' an increase in NAD+. When NAD+ increases, DNA repair mechanisms improve. We saw that in Chapter 2.

Sirtuins also help control 'catabolic' metabolism. Catabolic means the necessary deconstruction of cells to provide the material for new cells.

In this way, increasing the activity of sirtuins mimics the effect of calorie restriction and creates a conservation effect that promotes healthy ageing and increased longevity – at least in mice.

Other activators of sirtuins include polyphenols like resveratrol and quercetin, olive oil, green tea, turmeric, vitamin B3 as nicotinamide, soy isoflavones and possibly omega-3 – essentially the same foods that activate AMPK. Metformin is also thought to be a sirtuin activator, as is aspirin.

Chapter SUMMARY

This has <u>definitely</u> been a challenging chapter! But it is central to delaying ageing. Here is the summary.

- Turning <u>down</u> the insulin sensing (IGF-1) and mTOR pathways promotes healthy longevity.

DECREASE
INSULIN SENSING

DECREASE
MTOR ACTIVITY

Turn DOWN IGF1 and mTOR

- Conversely, turning <u>up</u> the activity of AMPK and sirtuins equally supports healthy longevity.

INCREASE
AMPK ACTIVITY

INCREASE
SIRTUIN ACTIVITY

Turn UP AMPK and Sirtuins

- All four nutrient sensing pathways increase autophagy – meaning senescent cells are cleared away and consumed before they contaminate other healthy cells with toxins.

Not only does autophagy remove these zombie cells before they do harm, but it breaks down the cells to be re-used as parts of new healthy cells.

It's a recycling process that is so important that leading healthy ageing experts from Harvard and University College London concluded:

"Enhancement of the autophagy process is a common characteristic of all principal ... anti-ageing interventions." [92]

The ideal healthy ageing strategy, therefore, would be to improve the functioning of all four nutrient sensing pathways.

- Remember the Blue Zones and where they are? You'll see that the Mediterranean diet contains polyphenol rich fruits and vegetables, soluble fibre, plant proteins, red wine and olive oil.

- The Asian diet is high in isoflavones from soy and in green tea. In these societies many people stay physically active into old age. All these elements improve nutrient sensing.

- The relationship between challenge and success in humans is striking. An above average number of successful entrepreneurs, like Richard Branson, have dyslexia[93]. And many others had a difficult early home life or economic disadvantage. The challenges brought out a determined response and were a critical factor in their later success.

- In a parallel way, occasional challenges to the body prompt positive responses that we now know increase health and longevity. That's why occasional hot or cold shocks (think a cold shower), fewer calories and exercise all work to modify nutrient sensing in a way that improves health.

- 8 -

Get your cells talking to each other

You need your cells to be chatterboxes. Cells must communicate continuously with each other through chemical signals to maintain health.

For example, when your pancreas detects you have eaten, we have seen that it releases the hormone insulin to tell other cells in the body to remove glucose from the blood. If that signal is confused or doesn't reach its target, blood sugar rises to toxic levels, creating a pre-diabetic condition.

We saw in the last chapter that cell communication is fundamental to the way the body responds to its environment, to nutrition and activity levels. Clear cell communication is central to the four nutrient-sensing pathways which are, in turn, central to extending health-span.

Cells also need to sense each other's boundaries, otherwise the integrity of the cell can be compromised. If that happens, cells division can become faulty and cell growth becomes uncontrolled – a hallmark of cancer.

Cell miscommunication is involved in multiple sclerosis – a disease in which the protective sheath around nerve cells in the brain is damaged. Consequently, affected nerve cells can no longer transmit signals clearly from one region of the brain to another. Poor cell communication is also deeply involved in other auto-immune diseases like type 1 diabetes, IBD and psoriasis.

Your immune system equally needs to respond quickly to a signal that a foreign pathogen has entered the bloodstream, otherwise infectious disease can take hold. But that ability declines with age.

In Chapter 2, we quoted Professor David Sinclair of Harvard who defined ageing as:

> *"… a loss of information as cells are copied over and over throughout life."*

It's a bit like the child's game when you whisper a sentence to someone, who whispers it to someone else, who whispers it etc, etc. Eventually the original message gets scrambled and corrupted.

In a similar way, when intercellular or intracellular messages become blocked or scrambled, the result is damage and ageing is accelerated. In fact, most diseases involve at least one breakdown in cell communication.

That's why poor 'cell signalling' is yet another *Hallmark of Ageing*. So, we need to restore cell communication.

When cells fail to communicate

In the original paper on *The Hallmarks of Aging*[1], the authors list what causes the failure of cells to communicate properly.

A prime cause is 'inflamm-ageing', the persistent, low-level inflammation that is so central to ageing and which in turn has multiple causes. These causes include:

- The failure of senescent cells to be cleared away.

- Elevated blood sugar levels, type 2 diabetes and obesity.

- Blocked stem cell production

- A decline in immune function, which then fails to clear infectious agents, which then triggers more inflammation.

- A natural, age-related decline in growth hormone – which also leads to bone fragility, muscle weakness, low energy levels and the reduction in the growth of new neurons.

- A reduction in sirtuin activity

You can see from this list how so many of the hallmarks of ageing are interconnected. But we have also seen how the right foods, nutrients, increased activity and stress reduction can counteract so many of the threats.

Human growth hormone (HGH)

Human growth hormone (HGH), as the name implies, it is most influential in the early years and starts to decline after about 30.

The hormone is released from the pituitary gland in short bursts, usually after exercise and during sleep. HGH indirectly affects your metabolism and works with another hormone we have encountered, called IGF-1, to help build bone and muscle. So lower levels imply an increase in body fat and less muscle.

A good 7-8 hours of restful sleep is essential to optimise your levels of HGH[94][95]. Adequate protein intake and exercise are also needed. It appears that intense exercise produces the biggest bursts of HGH[96], as does a hot sauna session[97].

In yet another example of how so much is connected, we now know that intermittent fasting helps raise Human Growth Hormone. Why? Because your blood sugar levels are low during that period, which means insulin is low and that prompts the release of HGH.

Restoring cell communication

One of the main causes of scrambled cell communication is a problem we have already encountered – cell senescence. You will remember that if cells fail to die off completely and be re-purposed into new cells, they become senescent or zombie cells.

At this point, they spew out inflammatory toxins that prompt neighbouring cells to become senescent. A process called 'inflamm-ageing'. In turn, inflamm-ageing is a key cause of muscle wasting, heart disease and stroke, and is strongly implicated in Alzheimer's.

Therefore, one of the best ways to restore and maintain clear cell communication is to clear out senescent cells with senolytics and autophagy, using the strategies we discovered in Chapter 1.

The original authors of the *Hallmarks of Aging* paper suggest that, since inflammation is a key driver of poor cell communication, then you should consider taking a daily low-dose aspirin.

A paper in the journal *Cell* has the intriguing title "Aging, Rejuvenation and Epigenetic Reprogramming: Resetting the Aging Clock"[98]. It details how the accumulation of DNA damage within cells is also a factor in cells failing to communicate properly. The good news is that we have seen (in Chapter 2) how it is possible to boost natural DNA repair with a range of plant nutrients and polyphenols.

Uninterrupted cell communication depends on a supply of the full range of vitamins, minerals at optimum levels and a diet high in polyphenols. Omega-3 fats are especially important to create resilient cell membranes and they have a direct effect on keeping cell communication clear.

Choline and inositol are another two dietary compounds that are also important components of your cell membranes and support healthy cell communication.

Inositol helps transport signals across the membranes of your cells. It is found in beans, wholegrains and the bran of grains like wheat or brown rice. Higher intakes of inositol have been linked to a lower risk of cancers, especially colon cancer.

Choline is a necessary component to make a vital form of lipid (fat) in your cell membranes called phospholipids. Good dietary sources of choline include eggs, meat, poultry, fish, peanuts, and dairy products.

Since your cells use proteins as messenger molecules in their communications, the quality of the protein you eat is important in supporting healthy cell membranes and in cell signalling.

Finally, we have seen how vital it is for your body to be able to sense the quality and amount of nutrients available to it – cell signalling is the medium through which that happens.

It's all connected!

Chapter SUMMARY

- Improving the way cells communicates depends on a full range of vitamins and minerals at optimum levels.

- Once again plant polyphenols are vital, as is omega-3 and you should look to ensure good levels of choline and inositol in your diet.

- Consider, too, a low-dose aspirin if your doctor approves.

- 9 -

'Shape-up' your body proteins

We normally think of proteins as part of our food. But our bodies make thousands of different types of proteins. Proteins do most of the work in your body. They transmit signals, move oxygen around the body, create structures like collagen, create immune antibodies, and read the genetic code stored in DNA.

A protein's function depends on its unique shape, and it folds to create that shape. So, if proteins become misfolded or misshapen, they cannot function properly. Organs malfunction, brain activity is compromised, bones weaken, and immune function declines.

Unstable proteins are the ninth Hallmark of Ageing

Unfortunately, protein formation is a complex, error-prone process. There is a mechanism in the body that tries to keep the production of proteins stable and without errors or defects. This mechanism uses what are rather charmingly called 'chaperones'. Chaperones check the quality of proteins and, if necessary, refold them in the correct way.

Over time, however, this mechanism accumulates errors itself and your body can produce too few or too many proteins. They may also become misfolded; literally bent out of shape.

Dysfunctional proteins are behind cystic fibrosis, Parkinson's disease, cataracts and are the tau protein tangles and amyloid plaques in Alzheimer's disease.

What triggers these errors? You won't be surprised to know that DNA mutations, free radical damage, cell signalling errors, and failure of the autophagy process to clear senescent cells are all involved.

Environmental toxins and inadequate nutrition are also factors[99].

Counteracting unstable proteins

To fix misfolded proteins, we either need to:

- Prevent proteins becoming misfolded in the first place

- OR support the chaperone refolding operation

- OR activate the clearance of defective proteins ie. prompt autophagy

- OR counter the emotional and environmental stressors that are toxic to proteins.

Preventing and reducing protein errors

Comprehensive reviews of how natural food compounds can help prevent protein errors have been conducted at the University of Florence in 2013[100] and again in 2018 at the University of Campinas Sao Paulo[101].

These meta-studies identified polyphenols – especially those listed below – as a natural and potent defence against both protein misfolding and the clumping together of faulty proteins.

This clumping together involves amyloids, a sticky accumulation of proteins. It is amyloid plaques in Alzheimer's that collect between brain cells and disrupt cell function. Far less known is the fact that amyloid plaques can form in the heart – a cause of heart disease.

Amyloid proteins present another danger – once clumped together they excrete toxins that further damage tissues, especially in the brain – a process called neurotoxicity.

Key polyphenols have been identified as helping keep proteins normal.

Green tea

EGCG or epigallocatechin-3-gallate (why <u>do</u> scientists use such mouthful names?) is the most abundant polyphenol in green tea. It helps both to prevent amyloid formation and it also binds to misfolded proteins so they can be eliminated from the body.

In order to maximise green tea polyphenols in your system, the University of Florence study defined the ideal time to take green tea extract. It is after an overnight fast and together with vitamin C and 1,000 mg fish oil for omega-3 fatty acids. These appear to enhance the bio-availablity and effectiveness of EGCG[102].

A study in elderly Japanese subjects showed that the higher the green tea consumption, the lower the rate of cognitive impairment.

Omega-3

Other studies have confirmed that omega-3 is very important to protect the fatty membranes surrounding brain cells – neurons.

Although omega-3 from fish oil is better metabolised, you can also get a form of omega-3 called alpha linolenic acid from nuts (walnuts especially), chia seeds and flaxseeds.

Curcumin

Curcumin is the active ingredient in turmeric and has anti-inflammatory, antioxidant, and cancer preventive properties.

A recent study at Michigan State University shows that curcumin also supports the normal folding of proteins and can also prevent their clumping together[103]. This agglomeration, as other researchers remind us *"is the first step in Parkinson's disease."* [104]

Population statistics support curcumin for brain health. People in India have a 75% lower age-related incidence of Alzheimer's compared to the US or UK. Of course, curry uses turmeric and curry consumption in old age has been associated to better cognitive function. So, get stuck into that tikka masala!

Grapeseed extract

Grapeseed extract may sound like a rather fringe dietary supplement. In fact, grapeseed extract is a rich source of polyphenols – including resveratrol – and there are hundreds of studies to support it as a component of a healthy ageing programme[105] [106].

Grapeseed's high polyphenol content means it is both an anti-inflammatory and antioxidant, helping to reduce the oxidation of LDL cholesterol (the bad form). That's an important protection, because oxidation of LDL cholesterol plays a central role in atherosclerosis, or the build-up of fatty plaque in your arteries[107] [108] [109].

Grapeseed extract has also been shown to help improve blood flow and therefore lower blood pressure. A recent study showed it may even play a role in protecting against colorectal cancer[110].

Grapeseed extract is high in two important brain-protecting polyphenols – resveratrol and gallic acid. Both help prevent the aggregation, or clumping together, of proteins. Since, as we've seen, clumps of mis-formed proteins emit toxins, it therefore helps reduce toxicity in the brain[111] [112].

Quercetin

This is another polyphenol found in many foods of vegetable origin, including tea, onions, garlic, apples, cherries, cocoa and red wine.

In a study carried out on 39 flavonoids, quercetin was found to have a particularly strong ability to inhibit the clumping of beta amyloid. Fisetin had a similar effect – remember how the fisetin in strawberries had a powerful effect on clearing out 'zombie cells? Quercetin is also known to activate the AMPK signalling pathway[113].

Olive oil – extra virgin

This contains many beneficial polyphenols, two of which of have been investigated for their ability to prevent amyloid aggregation and its toxic effects both in vitro and in vivo[114 115].

The University of Florence meta-study on polyphenols and healthy brain ageing concluded that dietary supplementation of one of the olive oil polyphenols (oleuropein) in mice *"strongly improved their cognitive performance"*; that plaque deposits were *"remarkably reduced"* and an *"astonishingly intense autophagic reaction"* was triggered[116].

Add some herbs

Health researchers are including several herbs in their search for natural compounds that help prevent protein folding errors – and hence boost healthy longevity. They include rosemary, oregano, sage, thyme and peppermint.

It will not have escaped your attention that the Mediterranean diet is high in extra virgin olive oil, red wine, spices, berries and aromatic herbs, especially rosemary. And all the Blue Zone diets are high in polyphenols.

Supporting those protein chaperones

We have seen that chaperones are the mechanism whereby mis-folded proteins could be normalised[117 118]. Heat shock proteins are essentially chaperones; they look for damaged proteins and help to ensure that those proteins are rapidly recycled[101].

Certain events challenge our cells – including moderate-intensity exercise, or temporary calorie restriction, or exposure to cold or heat[119]. In response to these challenges, heat shock protein activity increases for a while to compensate. As we have noted, a challenge in biology, as in life, can bring out the best in you.

Since ageing is basically an accumulation of unrepaired damage, you would expect that boosting heat shock protein activity and therefore protein repair, should translate into longer health span. That is indeed true in animal models and population studies confirm it in humans.

So do Finnish people, who take regular saunas, live longer? Intriguingly, a 2015 article in the *Journal of the American Medical Association* suggests they do[120].

They conclude:

> *"Increased frequency of sauna bathing is associated with a reduced risk of coronary heart diseases and all-cause mortality".*

Conversely, a high fat diet and obesity both reduce heat shock protein activation.

People who exercise more or eat less, providing they still obtain optimum micro-nutrient levels, have more effective self-repair systems, and they live longer.

The inside-out way to look younger

How does your body (mostly!) keep its shape?

That's the role of the extra-cellular matrix or ECM. This is a network made up of collagen, elastin and other fibres which form outside cells. You can visualise it like a mesh or netting.

The vital role of collagen

Collagen is the most abundant protein in the body and a key component in skin, hair, nails, the cornea of the eye, the lining of blood vessels and other connective tissue in the body.

The word collagen is derived from a Greek word 'kolla' meaning 'glue', so collagen is part of the 'glue' that holds the body together.

But collagen production declines from about age 30. That decline accelerates in both men and in women after the menopause and with smoking and UV rays – photo-ageing. The visual result is wrinkles, less elastic skin and less water binding properties of the skin. All of which has led to a plethora of anti-ageing skin creams and products, aka 'hope in a jar'.

A skin cream can temporarily moisturise the outside of the skin, but researchers generally agree that collagen molecules are too large to penetrate the skin. So, creams with collagen will not increase collagen levels.

Moreover, as collagen is a protein, it cannot be digested whole – it needs to be broken down in the stomach like all other proteins we eat. So, despite the claims of some products, you cannot eat collagen to build collagen.

However, you can expect that taking action to reduce protein errors would help slow collagen loss – because collagen is a protein. Therefore, it should, literally, show up in better skin tone.

This inside-out approach to maintaining skin tone and keeping skin younger-looking can be further improved.

First, you can help reduce free radical damage to the extra-cellular matrix and skin with a diet that features antioxidant foods and nutrients like mixed carotenoids.

Second, you can reduce glycation. Remember that glycation is the combining of sugars in the blood with protein – resulting in the formation of advanced glycation end-products or AGEs. These contribute to the cross-linking of protein fibres and therefore stiffness and inflexibility. This not only happens in the skin, but in arteries, meaning glycation is also a heart disease risk.

But there is even more we need to do to protect the ECM and skin – and in doing so you will be reducing the risk of cancer.

MMPs and the ECM

Carlos López-Otín, who headed the team which wrote the seminal paper *The Hallmarks of Aging*, is a co-author of subsequent 2017 paper – "The role of matrix metalloproteases in aging"[121].

A protease is an enzyme that increases the rate of breakdown of proteins. Matrix metalloproteinases or MMPs are especially dangerous enzymes. They contribute to ageing by damaging tissue in the extra cellular matrix (ECM) – the external result being skin damage.

MMPs also adversely affect stem cells and hasten the development of senescent cells. This contributes to the ageing of the body and the development of neurodegenerative diseases.

If that is not bad enough, MMPs have a direct role to play in arthritis, cardiovascular disease and have a vital role in cancer.

If cancer cells remain dormant, they are only a potential danger. It is when they spread – metastasise – that they become life-threatening. Cancer cells also (indirectly) produce MMPs and by degrading the extra-cellular matrix – essentially chewing holes in it – MMPs break the barrier holding back cancer cells and allow them to progress and spread[122][123].

So, we need to protect the extra-cellular matrix from MMPs. That is the aim of much on-going drug research[124] – but natural compounds can definitely help.

Counteracting the danger of MMPs

There is not just one type of MMP. They are a family of at least 26 members.

Studies[125] [126] have shown that some of the most effective MMP inhibitors are curcumin, green tea extract, genistein (a key nutrient in soy isoflavones), resveratrol, nicotinamide, oleanolic acid, (found mainly in olives, olive leaves and bilberries), omega-3 fish oil and several marine polyphenols found in brown seaweed[127].

Certain of these natural compounds will inhibit some types of MMPs better than others, so these nutrients act best synergistically, and consequently they should be combined.

One group of MMP inhibitors – the procyanidin flavonoids that include grapeseed and bilberry extract – appears to act as a shield wrapping around the extra-cellular matrix fibres to protect them.

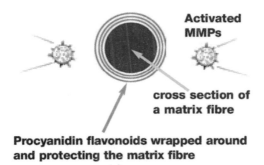

Activated MMPs

cross section of
a matrix fibre

Procyanidin flavonoids wrapped around
and protecting the matrix fibre

Collagen, to repeat, is a protein found in nails, hair and the cornea. So, a comprehensive programme to inhibit MMPs should not only help protect eyesight – it should show up as an improvement in hair and nails.

This is often what people notice first when they use our recommended *Delay Ageing* plan.

It is the outside confirmation of inside results.

Supporting the extra cellular matrix

Maintaining the internal structure of the ECM is not just a matter of protecting it from free radical, glycation and MMP attack. The ECM, like all tissue, is constantly subject to wear and needs repair. But if the rate of wear exceeds the rate of repair, then it deteriorates.

Improving the rate of repair depends on a full range of vitamins, minerals, polyphenols and amino acids from protein in the diet. It also needs an adequate supply of amino sugars, of which the most important is glucosamine. The body can make glucosamine, but the rate of production declines with age – with one result being slower wound healing.

Therefore, any programme to reduce skin and ECM ageing – and to delay ageing generally – should include supplementing with glucosamine. That means it must also include vitamins D3 and K, which are co-factors to ensure glucosamine is properly utilised, plus vitamin C and zinc for collagen and elastin synthesis.

Most people are familiar with glucosamine as a supplement to help in cases of arthritis. Arthritis is another problem caused by wear exceeding repair.

Finally … proteins in the brain

The brain is especially vulnerable to oxidative stress and free radical damage because it is over 60% fat. The brain also consumes a lot of oxygen. Although it is only 3% of your body weight, it uses

20% of your oxygen intake. It equally uses a lot of glucose, and you will remember that burning oxygen and fat creates free radicals.

So, reducing free radical damage, through antioxidant flavonoids, increasing cell repair mechanisms and reducing protein folding errors are all important ways to protect the brain and delay brain ageing.

Chapter SUMMARY

- Keeping your cells making healthy proteins relies on a full range of vitamins and minerals. Especially protective seem to be flavonoids in berry fruits, in sulphur foods like onions and garlic, the EGCG flavonoid in green tea, grapeseed extract, curcumin and olive oil.

- Protective foods include flaxseeds, chia seeds, garlic onions and dark chocolate – together with several herbs that include rosemary, sage and oregano.

- Delay the ageing of your ECM – and therefore skin – by supporting collagen formation with vitamin C and zinc and improving the rate of repair with glucosamine.

- Defend your ECM from MMP attack with polyphenols that include curcumin, green tea extract, grapeseed extract, omega-3, nicotinamide and resveratrol.

- 10 -

Re-balance your second brain – the gut

You are less than half human – which is a fact not an insult!

Of the estimated 80 trillion total cells in the human body, about 37 trillion are human and about 40 billion are microbial.

Your microbiome is the vast collection of micro-organisms that co-habit with us. They are mostly bacteria and yeasts and largely in the gut.

The trillions of tiny benevolent 'good' bacteria that live in our lower intestines metabolise the food that the stomach will have partially digested, extracting the nutrients we need.

Weighing a total of about 5 lb or 2 kg (more than 2% of your total weight!), they help to defend us against and 'crowd out' bad (pathogenic) bacteria which can otherwise produce toxins that are absorbed into the bloodstream and cause disease.

Those microbes and their genes have an enormous impact on the rate at which we age and on our health.

For example, the composition of microbes and bacteria in your gut has a big influence on how easy (or difficult) it is to stay slim, on whether you develop type 2 diabetes, or asthma, whether you suffer from gut inflammatory diseases like IBS, and whether you can better resist candida and thrush infections.

The range and type of the microbes in your gut also influence your ability to challenge and recover from infection, as 70% of your immune system is controlled by the gut[128].

Just as importantly, the fact that there is a continuous two-way communication between the gut and the brain means that gut health is now known to strongly influence mood and is a factor in depression[129].

Gut bacteria also create vitamins (like vitamin K, B12, folate and biotin) and amino acids – and they directly dial up the expression of genes that can supress inflammation and improve your immune response to the threat of colorectal cancer.

I trust we are making a good case for adding Gut Health as the tenth marker of healthy ageing! You need to become a 'clever guts'.

Your 'second brain' is in your intestines

Researchers call the enteric nervous system in your gut 'your second brain'. It comprises about 500 million neurons (nerve cells) lining the long tube of the gastrointestinal system. That's about as many neurons as in the brain of a labrador dog.

We now know that the gut also produces hormones and chemically communicates directly to the brain – conveying emotions and affecting mood. That connection or 'hotline' between the brain and the gut is via the vagus nerves.

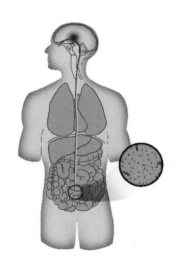

'Butterflies in the stomach' are part of our stress response originating in this second brain. A 'gut feel' that we don't trust someone is literally our brain below communicating to our brain above. And we've all eaten something that 'didn't agree with us' – and later our gut brain 'knew' to avoid that same food again.

The enteric nervous system (ENS) uses over 35 neurotransmitters in a similar way to the brain and an extraordinary 90% of the body's serotonin is made in the gut[130].

Serotonin is the 'feel-good' chemical that nerve cells produce and is responsible for maintaining a stable mood balance – a sense of wellbeing. Researchers are finding that an alteration in the balance between 'good' and 'bad' intestinal bacteria can directly alter serotonin levels, mood, anxiety and even confidence[131].

Accentuate the positive (bacteria)

There is a daily battle going on inside you, good versus bad.

The key to gut health is to ensure your gut has a favourable ratio of good 'friendly' bacteria to bad or pathogenic, disease-causing microbes. The role of the 'good' bacteria, called probiotics, is to counterbalance the bad, crowding them out and weakening their influence.

A poor ratio can allow infections like salmonella and E. coli to take hold. An overgrowth of pathogenic microbes is linked to frailty, inflammation, leaky gut (a very real syndrome), colorectal cancer and brain diseases like Alzheimer's and Parkinson's.

You also need to ensure you have a wide range of probiotics in your gut because microbial diversity is a marker of health – but normally declines with age.

In 2019, researchers at the University of Arkansas surveyed the literature on gut health and longevity. They highlighted a study on healthy, long-lived Chinese which showed that their microbiomes were more diverse than their sicker, same-age compatriots[132].

Significantly, their guts were enriched with bacteria that creates a fatty acid called butyrate.

The role of butyrate

Butyrate is created when our gut bacteria digest tough plant fibres and turn them – by fermentation – into butyrate fatty acid. This supports the immune system and helps protect you against diseases like colon cancer.

Butyrate also comprises 90% of the fuel used by the cells lining your gut. If the gut wall lacks integrity, a result is a 'leaky gut'. This is a condition where microscopic holes in the gut wall allow toxins and semi-digested food particles to enter the bloodstream – triggering inflammation.

Gut inflammatory diseases include IBS, Crohn's disease and colitis, and gut inflammation is linked to depression and dementia[133].

The importance of gut diversity is reinforced by two other studies. A cohort of long-lived healthy Italians were found to have a much wider range of healthy probiotic bacteria in their intestines than a matched panel of older, but less healthy Italians.

The same results were found in Japan. Increased microbe diversity and butyrate forming bacteria were linked to healthy longevity.

Different national diets – but the same pattern.

The Arkansas researchers concluded:

> "… *modulation of the gut microbiome to maintain a healthy gut will promote healthy aging. We further hypothesize that modulation of the disturbed gut microbiome to a healthy one, in the elderly with chronic diseases, will alleviate their symptoms and increase the quality of their lives.*" [132]

So, creating a high ratio of beneficial bacteria to bad microbes and encouraging butyrate is yet another key to healthy ageing.

Health researchers suggest that the balance of gut flora should be approximately 85 percent good bacteria and 15 percent other bacteria. If this ratio becomes unbalanced – a condition known as dysbiosis – the result is a compromised immune system and ill health.

But first, let us look at what can cause your gut microbes to become unbalanced in the first place.

The causes of an unbalanced gut

A poor diet

The key point to remember is that gut bacteria thrive on what you feed them. They eat what you eat!

A nutrient-poor diet, and sugar in particular, promote unhealthy bacteria and yeasts in the gut[134]. And a diet that has too few plant foods to provide fibre for butyrate-forming bacteria to flourish will encourage the proliferation of harmful bacteria.

Other causes of gut dysbiosis include some artificial sweeteners and pesticide residue.

Antibiotic treatments

One reason why beneficial bacteria decline is the use of antibiotics. Although the development of antibiotics is amongst the greatest medical advances of the 20th century, and a key cause of increased lifespan, antibiotics have a flaw. They kill pathogenic bacteria, but they also attack good bacteria along the way – internal 'friendly fire'.

A course of antibiotics causes a reduction in the number and range of protective probiotic bacteria, which is why antibiotics can cause nausea and upset stomachs and encourage candida yeast infections. The imbalance can last up to a year or more.

By killing good as well as 'bad' bacteria, antibiotics can also cause gastrointestinal inflammation and vulnerability to pathogens like C. difficile. Salmonella can cause infection at a dose 1,000 times lower if the patient is on antibiotics.

Therefore, you should only take an antibiotic when absolutely necessary, and always complete the dose. If you do not complete the course, it is quite possible that some of the original pathogens will be left and they can, and will, develop resistance.

One of the most dangerous past advertising slogans was "Kills 99% of all germs". It is precisely the 1% left that could mutate to become resistant!

That's because bacteria reproduce and evolve at a staggering rate. Given the right conditions, a single bacterium like E. coli can multiply exponentially to become a colony of 16 million in eight hours[135].

Over-prescription of antibiotics is one reason we are seeing a massive evolutionary change in bacteria – the rise of the super-bugs that make a visit to a hospital a risky trip[136].

Those of you with young children or grandchildren should note that Dr Martin Blaser of the New York University School of Medicine, and author of 'Missing Microbes', has a disturbing theory. He traces the link between high antibiotic use in childhood and the rise of food allergies, asthma, hay fever, juvenile diabetes and childhood obesity.

He argues that the over-use of antibiotics indiscriminately eliminates beneficial as well as harmful microbes – and thus adversely alters the composition and diversity of children's microbiomes. This directly leads to the rise of the health problems he highlights. We must, he warns, add back the *"missing microbes."* [137]

So, some of the dietary advice in this chapter very much applies to all ages, including children, as well as to ensuring healthy ageing.

Residual antibiotics in meat and milk

Unfortunately, it is not only a course of human antibiotics prescribed by a doctor that causes the problem. We also ingest antibiotics indirectly, as they can be residual in non-organic meats and milk. Indeed, the over-use of antibiotics in farmed animals is far greater than that in humans.

Stress

A decrease in healthy bacteria can also be caused by stress and elevated cortisol. Antacids and acid-lowering drugs against heartburn and indigestion also negatively alter the good/bad bacterial balance and can increase the level of disease, encouraging bacteria like C. difficile, which can cause severe illness.

Improve the balance in your microbiome

Consuming the right probiotic foods, and sometimes probiotic supplements, can help you bring the ratio of healthy bacteria back into balance. To over-simplify, more 'good' bacteria leave less room for the 'bad' ones.

Probiotic foods are essentially fermented foods. Fermentation occurs when the sugar in carbohydrates is converted into organic acids in an oxygen-free environment – your stomach. That acidic environment encourages the growth of beneficial probiotics and the suppression of pathogenic ones.

Eat more fermented foods

Every culture has had to rely on fermentation to preserve foods in the non-growing seasons:

- Germans developed sauerkraut

- Bulgarians made kefir and yogurt

- Asian cultures created tempeh, kimchi, natto (fermented soybeans) and miso

- Russians have 'raw' yogurt

- France and England have aged and blue cheeses

You can easily increase your intake of fermented probiotic foods by:

- Adding tempeh as a meat substitute in a stir-fry – but towards the end, as high heat can destroy active cultures

- Adding miso to vegetable or black bean soups

- Serving sauerkraut as a side dish

Although yogurt may seem an easy way to boost your intake of probiotics, the high-temperature food production process for mass market yogurts often destroys living bacteria. So, although good bacteria cultures are used in the initial process, by the time mass market yogurts reach the supermarket shelves, they may not contain many live and active probiotics. And flavoured yogurts are full of sugar.

Try probiotic supplements

The last five years or so have seen a huge rise in the research and sale of probiotic supplements. Strains have been developed that do survive the acidity of the stomach and reach the gut. We will look at the research in a moment.

Probiotics are critical to health – indeed the derivation is *pro* (for) and *biotic* (life). But probiotics on their own are not enough. Probiotics are living beneficial bacteria and like any living thing, they need their own food supply. These foods are called prebiotics.

Probiotics need prebiotics

Prebiotics are non-digestible plant fibres that bacteria break down and feed on. Prebiotic foods include garlic, onions, leeks, asparagus, bananas, dandelion greens, oats, barley and flaxseeds.

Many probiotic foods contain a lot of prebiotics, too. For instance, sauerkraut is based on cabbage, which contains a type of prebiotic which is also found in grains, legumes like beans, peas, chickpeas and soybeans and cruciferous vegetables.

The multiple benefits of a healthy gut

Better mental health

Recent studies have demonstrated that a healthy microbiome has a direct role in mental health, with people who regularly consume fermented foods and probiotics having less anxiety and depression[138] [139] [140] [141].

When the saliva of patients taking a probiotic in a recent study was tested, it was found to contain lower levels of cortisol – the stress hormone – than that of control patients.

Lower levels of anxiety may also be because the amino acid tryptophan is created during the internal fermentation, triggered by both prebiotics and probiotics.

Tryptophan synthesises serotonin, which alleviates anxiety. The sequence is:

PROBIOTICS→FERMENTATION→TRYPTOPHAN→SEROTONIN
→ **FEELGOOD FACTOR**

Recent research has shown that depression is frequently associated with gastrointestinal inflammation, because inflammation is rarely confined to one area.

There are cell signalling proteins (called cytokines) that act as chemical messengers to the body and brain. When cytokines are activated by a virus or bacterial attack in one area, they can call up a generalised but sometimes excessive immune response – called a cytokine storm.

That can then trigger inflammation in another area – in this case the brain – because we now know that cytokines can cross the blood-brain barrier.

This explains why a common side-effect of irritable bowel syndrome (IBS) is mild or even moderate depression. IBS is, of course, an inflammatory disease.

A 2017 issue of *Gastroenterology* reported on a small (44 person) group of people with anxiety and depression[142]. The group took a daily dose of the probiotic Bifidobacterium for 10 weeks. A control group took a placebo. At the end of the study, 64% of the probiotic group had lower depression scores compared with the control placebo group and improvements on a quality of life scale.

Any disease that ends in the suffix 'itis' means inflammatory. So, doctors who see patients with rheumatoid arthritis, or diverticulitis or even osteoarthritis, should be asking not just how they feel physically, but mentally. Everything is connected!

The direct link between inflammation in the body and depression in the mind is explored in an excellent recent book called 'The Inflamed Mind', by Edward Bullmore, Professor of Psychiatry at Cambridge University.

While not every depressed patient will suffer from inflammation, Dr Bullmore suggests that depressed patients should be tried on anti-inflammatory drugs or be encouraged in mindfulness training.

Since our priority is to use natural methods first, I propose a combination of mind calming, an anti-inflammatory diet and some powerful anti-inflammatory nutrients, possibly adding probiotics.

Reduced dementia symptoms?

Significantly, the characteristic plaques or tangles found in the brains of people with Alzheimer's are present in neurons in their guts, too.

A study published in *Frontiers in Aging Neuroscience* reported on 60 Alzheimer's patients who took a daily milk drink that included four probiotic bacteria species for 12 weeks[143].

The strains used were Lactobacillus (L.) acidophilus, L. casei, Bifidobacterium bifidum, and L. fermentum. There was a *"statistically significant"* improvement of cognitive function from the probiotic infused milk, compared with those who drank regular milk without probiotics.

Reduced risk of Parkinson's disease

Parkinson's is the second most common neurodegenerative disease. Researchers at Caltech University have confirmed that changes in the composition of gut bacteria are contributing to – or might actually cause – the deterioration in motor skills that characterises Parkinson's disease[144].

The initial clue was that 75 percent of people with Parkinson's have gastrointestinal (GI) abnormalities. These were primarily constipation and bloating. These GI problems often precede the motor symptoms by many years, says lead researcher, Sarkis Mazmanian.

Although the study was in mice – as inevitably so many preliminary studies are – the implications are that improving gut flora might be a way to cut the risk of this incurable disease. It is early days, however, as the researchers have yet to identify the best strains of probiotic that would form part of a preventative strategy.

Reduced stress

According to the World Health Organisation (WHO), stress is the fourth leading cause of disability worldwide. The 2015 survey 'Stress in America' worryingly indicates increasing levels of stress in adults – to the point where 78% reported experiencing at least one symptom of stress and 24% had incidents of extreme stress in a year.

There are now dozens of studies that confirm probiotics can have a role in lowering the damaging effects of stress, to the extent that several researchers now use the term 'psychobiotics'.

University Health News reported on a 2015 randomised, double-blind, placebo-controlled trial (the gold standard for human trials). The trial randomly assigned patients with major depressive disorder to receive either probiotic supplements or placebo. The probiotic consisted of 2 billion CFUs each of Lactobacillus acidophilus, L. casei, and Bifidobacterium bifidum.

After eight weeks, patients who received the probiotic had significantly decreased total scores on the Beck Depression Inventory, a widely used test to measure the severity of depression, compared with placebo[145].

In addition, they had significant decreases in systemic inflammation as measured by C. reactive protein (CRP) levels, significantly lower insulin levels, reduced insulin resistance, and a rise in glutathione, the body's key antioxidant.

We have already mentioned 'leaky gut syndrome', where the wall of the gut develops tiny gaps, but large enough for toxins and partly digested food to leak into the bloodstream. This powerfully stimulates inflammation throughout the body including the brain, which in turn can lead to depression, anxiety and impaired memory.

The gut–brain axis is so influential that scientists in the field are speculating that a deficiency of probiotics in older people contributes to memory loss and disorientation.

Improved immune function

A large-scale review of the role of probiotics in health published in the *American Journal of Human Nutrition* notes that:

> *"Allergic diseases, asthma, chronic inflammatory bowel disease, Crohn's disease, ulcerative colitis, diabetes, eczema, osteoporosis and arthritis are all strongly on the rise. These diseases arise from weakening of immune defense mechanisms in the gut."* [146]

Since 70% of the cells that make up your immune system are in your gut, your gut health has a big influence on the health of your immune system. In turn, the strength of your immune system determines how well you resist not just colds and flu, but long-developing diseases like cancer.

Regular probiotic supplementation has been shown to maintain intestinal health and enhance natural immune system response by stimulating the body's production of Natural Killer and T-cells[147].

Better digestion and nutrient absorption

There is evidence that, as we age, the body becomes less efficient in digesting food and extracting nutrients from it. A healthy level of 'friendly bacteria' can aid nutrient absorption by improving digestion. Bacteria do this by producing more of the enzymes that break down food. A small study showed that the probiotic strain Bacillus coagulans can also improve protein absorption[148].

By helping to increase the number of healthy bacteria in the gut, fermented foods and probiotics can also improve digestion and reduce bloating, gas, constipation and diarrhoea. But how?

Some strains like Lactobacillus acidophilus create lactic acid, which reduces the pH (acid/alkali level) of the intestine. This then speeds up the digestive process and allows stools to pass through the colon more quickly, in turn reducing the incidence of constipation.

Counter-intuitively, probiotics can also help stop diarrhoea. When the digestive system is overrun by harmful bacteria, the gut cannot absorb all the food – the result being diarrhoea. Fermented foods and probiotics that improve the good to bad ratio are therefore a potential natural alternative to over-the-counter remedies that can have unwanted side effects.

Lower LDL (the bad) cholesterol

A meta-analysis of 32 random controlled trials of probiotic supplements, published in the journal *Medicine* in 2018, concluded that they can *"significantly reduce"* total cholesterol by aiding the digestion of fat[149]. Some strains, like Bacillus coagulans, Bifidobacterium lactis and Lactobacillus plantarum, have been shown to help lower LDL (the 'bad' form of cholesterol) and increase HDL levels (the good form).

Creating natural antibiotics

We have seen that pharmaceutical antibiotics can negatively disturb probiotic balance. But some strains of probiotic can act as natural antibiotics.

The strain Lactobacillus acidophilus DDS-1 has been shown to produce the natural antibiotic-like substance called acidophilin, which can kill pathogenic bacteria like salmonella and E. coli.

Dr Khem Shahani, from the University of Nebraska, the developer of the L. acidophilus DDS-1 strain, has demonstrated that it can have a similar antibiotic effect to streptomycin (a strong antibiotic), but that the effect is selective – killing just pathogenic bacteria.

Dr Rob Knight leads a huge study on the microbiome called *American Gut*, and heads the Center for Microbiome Innovation at UC San Diego. He reports on a study with patients who had an acute C. difficile infection. One group of these patients was treated with antibiotics versus a matched group of patients who received a faecal transplant from a donor with a healthy microbiome. (Yes, a faecal transplant is exactly what it sounds like 😔).

The antibiotics were 30% effective. The transplant that created a more healthy microbiome was over 90% effective in clearing symptoms – within 36 hours![150] Other researchers have made similar discoveries[151].

Even a powerful multi-strain probiotic is very unlikely to work as fast as a faecal transplant, but over time could have a similar effect.

Reducing candida yeast infections

Candida yeast in normal amounts helps with digestion and the absorption of nutrients. But when candida yeast overproduces, symptoms may appear that include chronic fatigue, hormonal imbalances, mood disorders and thrush – which causes itching and discomfort.

Left unchecked, candida overgrowth breaks down the walls of the intestinal lining and penetrates into the bloodstream. This releases toxic by-products, causing leaky gut syndrome, and can even infect membranes around the heart or brain.

Researchers have identified sugar-rich Western diets, together with excessive use of antibiotics, as a reason for the big increase in candida yeast infections over the last few decades. Candida thrives on food sugars and vaginal candidiasis often follows antibiotic therapy. Other contributing factors include birth control pills and cortisone drugs.

Probiotics, especially the strains Lactobacillus acidophilus and Bifidobacterium bifidum, can stimulate the production of white blood cells that combat candida yeast and fungal infections.

A 2014 paper in the *Journal of Lower Genital Tract Disease* reviewed many clinical trials over a 20-year period. Most studies showed that Lactobacillus strain probiotics were able to significantly reduce the symptoms of bacterial vaginosis and reduce its recurrence. (Vaginosis is an infection caused by an imbalance of good and bad bacteria in the vagina.)

Even helping to resist cancer

We have seen that both prebiotic and probiotic foods can stimulate the production of butyrate – which is protective against especially colorectal cancer.

Nitrites used in food processing and especially in cured meats like bacon, salami and sausages can be converted to carcinogenic (cancer causing) nitrosamines in the digestive tract. Probiotics like L. acidophilus have been shown, in vitro, to help inhibit this chain of events[152].

Reduced osteoporosis

It is not generally appreciated that osteoporosis is a disease linked to changes in the immune system, as well as to calcium metabolism.

Multiple studies, summarised in a 2017 report at Michigan State University, showed that improving the composition of the microbiome via probiotics enhances the immune system and translates into improved bone density and reduced levels of osteoporosis[153].

This is partly because probiotics help the absorption of minerals required for healthy bone, including calcium, phosphorous and magnesium. And partly because improved gut balance improves the functioning of hormones that influence the building of bone. The report concludes:

> *"Modification of the gut microbiota, by ingesting probiotics, could be a viable therapeutic strategy to regulate bone re-modelling."*

A probiotic supplement following an antibiotic course is therefore especially important for women at risk of osteoporosis. The strains that seem to have the best effect include Lactobacillus casei, L. rhamnosus, L. acidophilus, and Bifidobacterium longum.

Another 2017 study indicates that combining probiotic supplementation with isoflavones can improve oestrogen metabolism and bone density in post-menopausal women[154].

Improved blood sugar levels

High blood sugar levels are part of a dangerous condition known as 'metabolic syndrome'.

Metabolic syndrome combines high blood sugar, high blood pressure, high cholesterol, and increased levels of adipose (body fat) tissue, especially around the tummy. This type of fat – called visceral fat – is especially dangerous. It releases toxic chemicals that further increase inflammation within body tissue.

Metabolic syndrome conditions very significantly increase your risk for stroke and heart attack, as well as diabetes. The *National Institutes of Health* has published several dozen studies showing that probiotic supplements, such as Lactobacillus and Bifidobacterium strains, can help reduce the risks for these diseases by helping maintain optimal blood glucose levels[155].

More effective weight loss and control?

It's a fact that normal-weight people generally age more healthily than obese people. While claims that probiotics may aid weight loss have been controversial, there is now increasing evidence that probiotics may contribute to normalising weight in conjunction with a healthy eating plan.

The *British Journal of Nutrition* noted that the intestinal flora of fatter people differs from thin ones. Could rebalancing their microbial status lead to weight loss?

To test this hypothesis, researchers recruited 125 overweight men and women. The subjects followed a 12-week weight-loss diet, followed by a further 12-week period aimed at maintaining body weight. Throughout the study, half the participants took 2 capsules daily containing probiotics from the Lactobacillus rhamnosus family, while the other half received a placebo.

At the end of the 12-week diet period, researchers observed an average weight loss of 4.4 kg (9 lb) in women in the probiotic group and 2.6 kg (6 lb) in the placebo group. During the next 12 weeks, the weight of the women in the placebo group remained the same, but the probiotic group continued to lose weight, up to a total of 5.2 kg (11 lb).

Therefore, women consuming probiotics lost twice as much weight over the 24-week period of the study. Researchers also pointed out that the appetite-regulating hormone leptin decreased, as did the intestinal bacteria that characterise obesity[156].

A 2015 meta-analysis in the *International Journal of Food Sciences and Nutrition* summarised 25 randomised human trials on the impact of probiotic consumption on body weight and BMI in over 1,900 healthy adults. They found taking probiotics did reduce Body Mass Index (BMI) and body weight[157].

Taking more than one type of probiotic for a period of 8 weeks or more resulted in the most weight loss. Other studies have shown that subjects who supplement with multi-strain probiotics can experience a measurable reduction in belly fat, the most dangerous kind of fat.

The potential for probiotics to contribute to achieving a healthy weight is logical, given the essential role of gut bacteria in breaking down food and nutrients. Another suggestion is that probiotics work via the gut-brain axis, increasing the release of a hormone called GLP-1 that reduces appetite and slows the absorption of dietary fat.

However, a probiotic on its own is not going to result in weight-loss – it is an adjunct, a booster, to a reduced calorie diet.

A diverse microbiome equals better health

The gut flora of Africans, South Americans, people following a Mediterranean diet and vegans are generally far more varied than those of most people in Europe or North America – and this is almost certainly due to the far higher levels of fibre in the diet. Early hunter-gatherers probably ate 10 times more fibre than we do now in the West. And far less sugar!

As we have seen, prebiotic fibre is food for probiotics and fermented by them in the gut to produce butyrate, which in turn lowers the risk of inflammatory bowel diseases, colon cancer, heart disease and diabetes.

Justin Sonnenburg, a Stanford University microbiologist, traces a clear path from low fibre diets to the range of modern health threats. He argues that our modern, fibre-deficient, Western diet means the 'good' bacteria in our gut have less to feed on. This results in less short-chain fatty acids like butyrate in the gut, and this, he states, leads to:

"... a simmering state of inflammation which is really the underlying cause of all the diseases like cancer, heart disease, allergies, asthma and inflammatory bowel disease".

Inevitably, Big Pharma is creating drugs to encourage the creation of butyrate. But why not simply increase your fibre content from vegetables, beans, fruits, nuts and whole grains and add some yogurt and fermented foods each week to your diet?

You can do one more major thing for gut health. The American Gut Project is a huge collaborative study and has already shown that the clearest factor in gut microbe diversity is the amount of exercise you do[158].

Better fitness and better gut health – two for the price of one! If you are American, you can actually join *The American Gut Project* humanfoodproject.com/americangut/. This lets you compare your gut microbes to those of others around the world.

The equivalent study in the UK is run by Professor Tim Spector of King's College London who directs the crowdfunded *British Gut Microbiome Project*. See britishgut.org/. Get involved; you will learn a lot and help push the science forward.

Chapter SUMMARY

- Think of those microbes in your intestines as a zoo filled with tiny, slightly unruly but mostly friendly, exotic creatures you need to feed for optimum health. Doing so helps the good crowd out the bad.

- The top priority is to ensure a good level of prebiotic fibre foods in your diet and fermented foods for the probiotics. The *Delay Ageing Food Plan* does that.

- I also recommend that you eat at least 20 different foods a week and try one new food a month. Why? Because if gut diversity is a sign of health, your gut needs a diversity of foods to feed those microbes. And new foods are fun!

- In common with many other health writers and researchers, I was initially sceptical about probiotic supplements. Did they survive the acidity of the stomach and reach the intestines alive? Did they really colonise?

  Some benefit claims for probiotic are weak – for example, there is little or no evidence that they can address tooth decay, eczema, or liver disease.

  I believe there is now good evidence that most people will benefit from taking a probiotic supplement after a course of antibiotics to help restore gut health. The same is true in cases of IBS and yeast infections – to help rebuild the ratio of good to bad bacteria.

- I also think there is enough evidence to occasionally try a multi-strain probiotic supplement to help improve the diversity of your gut and the ratio of good to bad bacteria. What is 'occasionally'? Perhaps two or three times a year for a month.

I say 'try', because different genomes will mean that some people will see a measurable benefit, others may not. But probiotics are not expensive, and change can be fast. Data from dozens of studies shows that the balance in your microbiome can be changed towards health in as little as two weeks from daily probiotic supplements – coupled with an increased intake of fibre.

- No-one really knows for sure the best strains for a multi-strain probiotic, but I believe it should include at least 8 or 9 of the following: Lactobacillus acidophilus, Bacillus coagulans, Bifidobacterium lactis, Lactobacillus casei, Lactobacillus rhamnosus, Lactobacillus plantarum , Bifidobacterium longum, Bifidobacterium bifidum, Lactobacillus gasseri and Streptococcus thermophilus. Different strains colonise in different parts of the gut and fulfil different roles.

- Look for a supplement with at least 10 billion CFUs or Colony Forming Units a day – or 300 billion over 30 days. 300 billion may sound like a lot, but it is still a fraction of the trillions of microbes in your gut. However, 'colony forming' means that the probiotic supplement is only designed to kick-start the development of beneficial colonies in your intestines.

- The level and range of probiotics above should not only help address inflammatory issues like IBS and leaky gut syndrome, but also candida. Research shows it might also have a beneficial effect for many people who suffer from depression and may complement and boost the results of a weight loss programme.

- 11 -

The immune system –
Fighting invisible enemies at your gate

During my lifetime (I was born in 1940), one of the proudest boasts of medical science had been that we had largely tamed infectious diseases. It is the principal reason behind the sharp rise in life expectancy during the 20th century.

Then we got the devastating awakening of COVID-19.

So, I don't believe a book on healthy ageing would be complete without examining how we can support the immune system. Especially as there is ample evidence that the immune systems weakens over time[159]. And this is linked to the increased risk of cancer, susceptibility to influenza and poorer response to vaccination.

We should not really have been so surprised at our vulnerability to the coronavirus, because there are literally tens of thousands of viruses and bacteria circulating in various parts of the world that could harm us.

Viruses and bacteria are emerging and mutating

They are emerging as road building, mining, logging, hunting, rapid urbanisation and population pressure push humans deeper and deeper into previously unexplored areas. Areas that harbour unknown pathogens that can jump from animals to humans – as the (reportedly) bat-transmitted coronavirus did in Wuhan.

David Quammen is author of the 2013 prophetic book 'Spillover: Animal Infections and the Next Pandemic'. He observes:

"We cut the trees; we kill the animals or cage them and send them to markets. We disrupt ecosystems, and we shake viruses loose from their natural hosts. When that happens, they need a new host. Often, we are it."

Kate Jones is Chair of Ecology and Biodiversity at University College London. In a Guardian March 2020 article, she describes transmission of disease from wildlife to humans as:

"… a hidden cost of human economic development. There are just so many more of us, in every environment. We are going into largely undisturbed places and being exposed more and more. We are creating habitats where viruses are transmitted more easily, and then we are surprised that we have new ones."

It is the disturbance of the local ecosphere that is also, almost certainly, behind the rise of Lyme Disease. And whilst the risk of new viruses jumping into the human population may be greatest in less developed countries, it can take just one infected passenger disembarking from an intercontinental jet into a crowded metropolis to spark a national or even global catastrophe.

Unfortunately, the problem is likely to get more acute. The most probable global temperature rise – predicted at 3 degrees Celsius – will make whole swathes of land virtually impossible to cultivate, in Africa especially. This will not only facilitate the spread of tropical diseases, but put pressure on food supplies, encouraging the continuance of 'wet markets' and triggering population movements.

In addition, higher levels of pollution weaken respiratory strength, which increases vulnerability to pathogens that are air transmitted.

At home, factory farms with huge herds of close packed animals are increasing the likelihood and virulence of bacterial threats and of flu viruses, which is what happened in the swine flu pandemic of 2009. A revised estimate of the death toll from that is now put at over 200,000 people[160].

Most British consumers are unaware that as many as 70% of farm animals may be kept on factory farms. The long-term human health cost of cheap meat is higher than most people imagine, quite apart from any concerns for animal welfare.

Previous flu pandemics

Industrial poultry production has been blamed for the emergence of the H5N1 bird flu virus. Whilst, fortunately, it is not easily transmittable to humans, the mortality rate is over 50%. The US Centers For Disease Control rate it, and H7N9, as our gravest potential flu pandemics if they mutate, as the latter virus has killed almost 40% of those infected.

We also have short memories. I lived through the Hong Kong H3N2 flu, where almost 80,000 people died in the UK between 1968 and 1970, when there was a second wave of the virus.

More on the H and N viral classifications

The 'H' in viral classifications stands for hemagglutinin and the 'N' for neuraminidase. These are both types of tiny protein spikes on the flu's surface that help it invade cells. Rather like suits or shoes, the spikes come in a variety of styles. There are 16 versions of hemagglutinin and nine of neuraminidase, and viruses shuffle the combinations as they mutate.

How viruses and bacteria spread

Bacteria are single-cell living microbes that generally infect us by contagion – contact. Bacterial infections include strep throat, urinary tract infections, bacterial food poisoning, Lyme disease and tetanus.

Viruses are far smaller and are not alive in the commonly accepted sense, as they can only replicate if they have a host. They hijack the host's internal cellular machinery to replicate. Viral infections include rhinovirus (colds), enteroviruses, herpes, adenovirus and of course coronaviruses. A virus is a particularly frightening threat, because you cannot see it, hear it or feel it.

It's time to know more about your immune system.

Your two immune systems

We are continually exposed to micro-organisms and toxins that could lead to illness. Whether they do or not, largely depends on the integrity of your two immune systems.

Your **innate immune system** includes the physical barrier of the skin and mucous membranes, plus specific cells that act as a generalised defence system that will react to new threats you have not encountered before. Think of these as a front-line, fast-mobilised patrol, continuously prowling the body looking for, and eliminating, <u>new</u> viral and bacterial threats.

Your second immune system is the **adaptive immune system**. In evolution, it came later. This is the one with 'memory'. It recognises viral or bacterial threats it has encountered <u>before</u> and mobilises immune cells to fight them. It's why you don't get the same cold twice.

Unfortunately, our immune systems normally get weaker as we age, and 'immuno-senescence' (ageing of the immune system) becomes a measurable issue from around 60, leaving this age group more prone to infection.

In a normal flu season, for example, very few people under 65 who contract flu get ill enough to be hospitalised. In contrast, about 20% of those over 65 do, and as many as 10% of them die. Immuno-senescence is also a reason why cancer is an age-related disease.

The good news is that researchers have discovered ways to slow, and possibly even reverse, the ageing of the immune system. The innate and the adaptive immune systems complement each other, but increasing the strength of your whole immune framework requires a somewhat different approach for each.

Supporting your ADAPTIVE immune system

When it is faced with a new pathogen like a toxin, bacterium or virus, your adaptive or acquired immune system creates specialised immune cells. These are T-cells, so called because they are created in the thymus, and B (lymphocyte) cells, which are created from bone marrow stem cells.

These T- and B- cells recognise the pathogen intruder as 'non-self' through molecules on its surface called antigens. After a few days, antibodies are created, which are specifically adapted to fight just that one viral or bacterial threat.

These highly specialised antibody cells 'remember' the threat and mobilise next time they are faced by the same pathogen. That's why someone who has recovered from measles is protected for life from getting measles again.

Although the response is far faster the second time that the adaptive immune system meets the same pathogen, it is clear that your Acquired Immune System, though very effective, has a chink. It will not immediately protect you from a new bacterial or viral threat. It requires time to build up immunity, which is why it takes days to get over a cold or flu caused by a new virus.

As most people get older, their immune response gets weaker, which is why they are prone to more infections, more flu, more pneumonia, more inflammatory diseases and, unfortunately, more cancer. And why they have a poorer response to vaccination.

This reduced immunity appears to be associated with a reduction in the production of T-cells in the thymus – a small central organ behind the breastbone. Its size normally shrinks from as early as your 20s and is replaced by fatty tissue.

Exercise can help produce more T-cells

However, a study on cyclists aged 57-80 at Birmingham University and King's College London has found that continuing exercise maintains immune function. Not only did the older cyclists' body fat levels stay similar to younger cyclists, but their thymuses were producing almost as many T-cells as the youngsters. The exercise had kept their immune systems from ageing.

Janet Lord, Director of the University of Birmingham's Institute of Inflammation and Ageing, is the study's lead author. She notes:

"Our findings debunk the assumption that ageing automatically makes us more frail."

But, of course, it doesn't happen without the exercise!

Regular exercise also supports the immune system by improving circulation. This allows immune cells to move more effectively through the body to where they are needed.

Micro-nutrients support the immune system

Most research, according to *Harvard Health*, points to 'micronutrient malnutrition' – depletion in essential vitamins and minerals – as being a key cause of immuno-senescence.

Deficiencies in vitamins A, B6, C, D3, E and folic acid (B9), and in the minerals selenium, magnesium, zinc and copper are known to lead to lower immunity[161].

Vitamin E

There are 8 versions of vitamin E. According to more than one study at Tufts University, the alpha tocopherol version of vitamin E enhances the innate immune system in such a way as to make it a suggested treatment in older individuals at risk for pulmonary (lung) infection[162]. Vitamin E supplementation at a level of 200 IU (133 mg) lowered the risk of acquiring upper respiratory infections in nursing home residents[162]. At that level it is safe, but higher levels should not be used.

Vitamin D

Recent research shows that the optimum level of vitamin D you need for healthy immune function, especially in the winter, is considerably higher than originally thought. It is now generally agreed at 800 IU or 20 micrograms throughout the year.

Although estimates vary, a review of the whole literature now indicates the truly effective dose for immune function is probably as high as 2,000 IU in the three winter months.

Top plant foods

Eating a largely plant-based diet will support your immune system, as it is plentiful in not just vitamins and minerals, but protective compounds like **beta carotene**, (which converts to Vitamin A as needed), **lutein** (which helps prevent the immune-lowering effect of high glucose in diabetics and pre-diabetics)[163] and a range of protective **flavonoids and polyphenols**.

Dark red, blue and black berries – including blueberries, blackcurrants and raspberries – are an especially good source of immune-enhancing polyphenols.

Citrus fruits should be included for their vitamin C content.

So should **green tea,** which contains a polyphenol called epigallocatechin-3-gallate (EGCG), which is shown to build immune function.

Garlic is a proven immune booster[164] and studies quoted by the National Cancer Institute indicate that people consuming six cloves a week or more have a 30% lower risk of colorectal cancer and a 50% lower risk of stomach cancer. So, add garlic and onions liberally to your cooking.

Leafy greens – broccoli is one of the best vegetables you can serve for immune support, but so too are cabbage, chard and spinach. Cook as little as possible to retain their power.

Nuts, seeds and avocados are good sources of vitamin E. As mentioned above, vitamin E not only helps damp down inflammation (inhibiting an inflammation pathway called COX-2), but also helps promote more activity in Natural Killer cells. This supports enhanced immune response and *"confers protection against several infectious diseases"* [165].

Shellfish contain zinc, which is an essential element in maintaining a protective immune function.

Mushrooms have been found to increase the level and activity of white blood cells and hence enhance the immune system – even to the extent that they show anti-cancer properties[166]. Look for oyster mushrooms in the supermarket and especially the less common shiitake, maitake and reishi, which are even more potent.

Combine them in **chicken soup**! An ideal combination would be to add garlic, onions and mushrooms to chicken soup. Nebraska University researchers have found that chicken soup contains a high level of the amino acid cysteine, which helps block immunity, lowering inflammatory cells. So, it's not a myth that chicken soup is good for immunity – as well as the soul.

Brown seaweed is common in many Asian diets and contains a compound called fucoidan. Recent in vitro research at the Sloan Kettering Cancer Centre suggests it can help reduce inflammation and improve the immune system[167]. Significantly, fucoidan may also reduce the damaging effect of viral infection on lungs, which occurs with Influenza A – a virus that has been responsible for at least three 20th century pandemics and coronaviral infections[168].

Stress reduction

Stress suppresses lymphocyte T- and B-cells, making your body vulnerable to illness and lengthening recovery times. So, make it a daily habit to carry out a simple, short, stress-reduction exercise like the one we detail later.

Supporting your INNATE immune system

Bacteria and viruses have an incredibly fast reproductive time – typically 20 to 30 minutes. Whereas, as we have seen, it takes your adaptive immune system a minimum of days, or up to weeks, to create future protective antibodies. The innate immune system protects you during the delay between microbe exposure and your first adaptive response.

The innate immune system includes protective barriers like the skin, respiratory tract, saliva, mucous and white blood cells. These white blood cells include dendritic cells, Natural Killer cells, neutrophils and macrophages. They recognise a pathogen as non-self and attack it.

This attack can involve spraying the intruder with deadly chemicals and, in the case of neutrophils and macrophages, literally engulfing and "eating" it (*macro phage* means big eater).

Statins to rejuvenate neutrophils?

The most common innate cells are neutrophils, but in an older body they seem to lose some of their ability to hunt down pathogens. Janet Lord, who led the cyclists' study on the thymus, muses that they seem to lose their sense of direction. In 2019, she was trying out existing drugs to see if neutrophils could be rejuvenated and was surprised to find that a very common drug appeared to do so. A statin.

Although this finding has yet to be confirmed in trials, it is true that people admitted to hospital with pneumonia are less likely to die if they are on statins. However, before we reach for the statins, which can have adverse side effects, there are other options.

Natural 1-3, 1-6 beta glucans

A natural compound called 1-3, 1-6 beta glucans can support and 'prime' the innate immune system. This compound is extracted from the cell walls of certain mushrooms or a type of baker's yeast, but in the latter case, the particles have been very finely milled and purified, and cannot trigger a yeast infection.

Over millennia, your innate immune system has learned to recognise yeast as a potential pathogen. So, when it senses the microscopic particles in the 1-3, 1-6 beta glucans (not the same beta glucans as those in oats), it responds by increasing the numbers and activity level of Natural Killer cells, neutrophils and macrophages.

Proven against viruses, RTIs, bacterial infections

This heightened activity results in more effective innate immunity, which then fights actual threats. These include flu or cold viruses, some respiratory tract infections (RTIs), TB (tuberculosis), pneumonia, and even bacterial infections like chlamydia, E. coli, salmonella or C. difficile.

A meta survey of over 20,000 other studies on various 'immune modulators' was conducted in 2019 at the Department of Immunology at the University of Louisville. It found that 1-3, 1-6 beta glucan has *"the best position"* among all other compounds tested. The researchers concluded:

> *"With more than 80 clinical trials evaluating their biological effects, the question is not if glucans will move from food supplement to widely accepted drug, but how soon … In addition, they are relatively inexpensive and possess extremely low risk of negative side effects."* [169]

Once again, we see the default position is to try to classify any effective nutritional product as a drug. Fortunately, 1-3, 1-6 beta glucans are not a drug, they are a natural compound.

Even positive effects on cancer and radiation damage

These studies have led other researchers[170] to successfully investigate the use of 1-3, 1-6 beta glucans in cases of cancer. They concluded that *"numerous animal and human studies have shown remarkable activity against a wide variety of tumors"* [171].

When Myra Patchen at the Armed Forces Radiobiology Research Institute in Bethesda tested 1-3, 1-6 beta glucans against even nuclear radiation damage, it was found to significantly improve survival[172] [173].

Beta glucans are also being used at the Brown Cancer Center in Kentucky and elsewhere as 'adjunct therapy' – therapy given in addition to primary cancer therapy to maximise its effectiveness[174].

A 2019 survey published in the *International Journal of Molecular Science* concluded that:

> *"… beta glucan may play an essential role in future strategies to prevent and inhibit tumor growth"* [170].

Safe and natural – cannot cause a cytokine storm

As a safe and natural molecule, 1-3, 1-6 beta glucans are overdue a much wider awareness and use.

Safety is an important issue, because the immune system can be over-stimulated. That can lead to what is called a cytokine storm – where the immune system becomes over-active and damages tissue. Whilst echinacea, for example, has an undoubted immuno-modulatory effect, some health writers have expressed concern that continuous or over-use of echinacea could lead to this situation. Beta glucans, however, <u>prime,</u> rather than stimulate, the immune system.

Probiotics can support BOTH immune systems

We saw earlier that some 70% of your immune system is located in the gut. This has encouraged research into the role of probiotics and probiotic supplements – 'friendly' beneficial bacteria.

A meta-analysis on probiotic supplementation conducted in *Nutrition Research* in 2019 found that probiotic supplementation over 3 to 12 weeks 'significantly' enhanced natural immune system markers in healthy older adults by stimulating the body's production of Natural Killer and T-cells[147].

Unfortunately, the survey had a limited amount to say on which strains seemed especially effective, but indications were that B. lactis, L. rhamnosus and L. casei are among the most effective. This conclusion is supported by other studies, which also point to L. plantarum and B. breve as having immune modulation benefits.

The IgA (Immunoglobulin A) antibody is a major component of the adaptive immune system. Antibodies are the proteins made by the immune system to fight antigens such as bacteria, viruses, and toxins. IgA is found in high concentrations in the mucous membranes, particularly those lining the respiratory passages, so it is important in combating flu.

A 2013 study[175] indicated that Lactobacillus strains may predominantly induce an innate immune response that helps kill tumour cells, whereas Bifidobacterium strains play a more regulatory role that involves the general priming of IgA T-cells.

Although probiotic research has been ongoing for some decades, there is still a lot we need to know, especially on the role of individual strains. However, a reasonable conclusion is that a multi-strain probiotic supplement is worth trying as you get older.

We know that it can help crowd out pathogenic microbes and helps you better digest nutrients that support the immune system. But equally important is to keep up your prebiotic fibre intake, to ensure that the probiotics have enough food to multiply and flourish.

Live yogurt can also help boost beneficial probiotic levels – but many brands in supermarkets are pasteurised, which can severely reduce positive probiotic activity.

Quorum sensing – don't underestimate viruses and bacteria

You may not have come across quorum sensing – but it has come across you!

A lone virus or bacterium is not going to do you any harm. It's only when they have increased their numbers to a critical mass that they pose a threat – and they appear to 'know' that.

Incredibly, viruses and bacteria can sense whether their population density is sufficient to mount a successful invasion; they have reached a 'quorum'. For that to happen, they send out chemical signals, called autoinducers, that express genes which then increase their concentration and, with it, their virulence and mobility.

In effect, they begin to act like a co-ordinated multi-cellular organism, much as a beehive consists of individuals, but works together. This coordination allows bacteria and viruses to wait until their numbers are large enough for a mass attack[176] [177].

Quorum sensing bacteria and viruses have a biofilm on their surface, containing sensors that achieve this remarkable communication. This is leading to the creation of drugs called 'quorum sensing inhibitors' that seek to disrupt the microbes' ability to send and receive signals. Meantime, some herbs and spices like ginger, garlic, cloves, vanilla and horseradish also have quorum sensing inhibitor properties.

In an age when it's well known that we are running out of antibiotics, this approach may give us hope of another route – a broad spectrum antibiotic based on inhibiting quorum sensing[178].

COVID-19 and immuno-senescence

Age is by far the biggest risk factor for a poor outcome from this new coronavirus. Not chronological age but accelerated biological age – accompanied by high levels of inflammation and a weaker immune system. So our recommendations could significantly reduce the risk of a severe result from contracting COVID-19.

Chapter SUMMARY

- Increase fruits and vegetables, including red berries, citrus fruits, broccoli, spinach, peppers, garlic.

- Ensure your diet includes mushrooms often – and ideally reishi, maitake and shiitake mushrooms.

- Eat at least 2 portions of whole grains a day.

- Eat natural, unsweetened, ideally organic live yogurt daily.

- Eat a small portion of mixed nuts and seeds a day – the vitamin E content supports the immune system.

- Take extra vitamin D in the winter – 2000 IU.

- Limit your alcohol consumption.

- Get at least 7-8 hours of sleep.

- Try a multi-strain probiotic supplement – certainly after antibiotics, which kill both good and bad bacteria.

- Eat more fermented foods – like sauerkraut, kefir, kimchi and kombucha. And fermentable fibre foods like oats, beans and legumes (peas, chickpeas and soybeans).

- Exercise helps maintain thymus activity.

- Take 1-3, 1-6 beta glucans, especially in the winter.

- mTOR inhibition can reduce 'immuno-senescence'. Reduce calories and eat more plant than animal proteins.

- And lose weight if you need to, as obesity negatively affects the immune system.

- 12 -

The story so far

HALLMARKS OF AGEING	HOW DO YOU COUNTERACT THEM? Foods, nutrients and activities to help neutralise the threat.
1 - Cells become 'senescent'	**Increase autophagy through senolytics** Eat more flavonoids in fruits and vegetables, especially fisetin in strawberries, mushrooms, peas and fermented foods, B3 as nicotinamide, spermidine in wheatgerm, grapeseed extract, carotenoids (lutein, lycopene and beta carotene), omega-3, genistein in soy. Exercise/Activity. Calorie restriction.
2 - Damage to DNA accumulates	**Boost DNA repair** Increase dietary antioxidants and boost SIRT proteins via NAD+. Polyphenols and flavonoids in fruits and vegetables, B3 as nicotinamide, vitamins D3, K and folic acid, selenium, zinc, carotenoids, green tea, curcumin, grapeseed. Exercise/Activity. Calorie restriction.

HALLMARKS OF AGEING continued	HOW DO YOU COUNTERACT THEM? Some foods, nutrients and activities to neutralise the threat.
3 - Mitochondria become dysfunctional	**Boost mitochondrial repair** Nicotinamide, vitamin E, B complex, selenium, zinc, CoQ10, omega-3, curcumin, green tea, flavonoids.
4 - Beneficial genes are switched off and harmful genes are on	**Trigger epigenetic change** Increase glutathione levels, so feature methyl donor sulphur-rich and methionine rich foods, folic acid, betaine, choline, flavonoids, and prebiotic fibre foods that activate butyrate – garlic, leeks, onions. Increase fermented foods. Exercise/Activity.
5 - Telomeres become shorter	**Slow down telomere loss** Omega-3, vitamin D3, folic acid, betaine, B12, olive oil, flavonoids, carotenoids, nuts, seeds. Exercise/Activity. Reduce stress levels.
6 - Stem cells became exhausted	**Reduce the loss of stem cells** Vitamin D3, berry polyphenols, green tea Calorie restriction. Exercise/Activity.

HALLMARKS OF AGEING continued	HOW DO YOU COUNTERACT THEM? Some foods, nutrients and activities to neutralise the threat.
7 - Nutritional sensing is weak	**Lower IGF-1 insulin sensing + mTOR Increase AMPK and sirtuin levels** Soluble fibre foods, polyphenol rich fruits (esp. berries), cruciferous vegetables, curcumin, omega-3, plant rather than animal protein, soy isoflavones. Exercise/Activity. Calorie restriction.
8 - Cells communicate poorly	**Reduce inflamm-ageing** Polyphenols, choline, inositol, omega-3. Low dose aspirin?
9 - Proteins accumulate errors	**Support chaperones** **Reduce protein folding errors** **Improve autophagy** Green tea, green tea extract, curcumin, quercetin, flaxseeds, grapeseed extract, olive oil, spices, garlic. Calorie restriction. Exercise/Activity.

OTHER HEALTH THREATS	HOW DO YOU COUNTERACT THEM? Some foods, nutrients and activities to neutralise the threat.
Microbiome becomes unbalanced	**Increase prebiotic/high fibre foods and try probiotics** Wholegrains, beans, peas, chickpeas, vegetables, fermented foods eg. sauerkraut, miso, kefir, yogurt. Try a multi-strain probiotic supplement.
Immune system weakens	Increase fruits and vegetables, especially red berries, citrus fruits, fermented foods and fermentable fibre foods like beans and legumes and mushrooms. Whole grains, nuts and seeds A vitamin D supplement in winter and all year for those with darker skins. Consider 1-3, 1-6 beta glucan supplement. Exercise/Activity.

AND Optimised full range of vitamins and minerals

All the hallmarks and other threats also need a full, optimised, range of all 23 essential vitamins and minerals to counterbalance them.

A Mind Map of the Story So Far

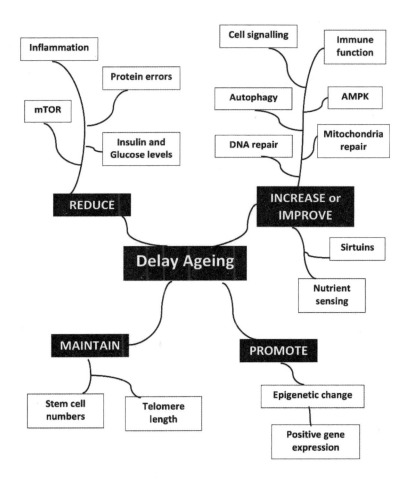

PART II

A COMPREHENSIVE PLAN TO DELAY AGEING

"HEALTHY TO A HUNDRED"

Ageing is largely the accumulation of damage[179]. Accumulative DNA damage, protein damage, mitochondrial damage, telomere damage, damage to cell communication. Wear and tear.

For the first three or four decades – our reproductive years – that wear and tear is repaired very well. Because natural selection has built our bodies to maximise the chance of us living long enough to reproduce and pass on our genes.

Once that's achieved, however, the genes that were selected for fertility are no longer so important. Nature starts to lose interest in you!

Therefore, without specific action, repair mechanisms begin to weaken. That is a common denominator of all the hallmarks of ageing. So, if you want to stay healthy, possibly for up to 20 years longer than average, your lifestyle needs to reduce wear and boost repair.

By targeting all 10 fundamental ageing factors, the aim is to delay or prevent the onset of all age-related diseases simultaneously, rather than wait to treat each illness individually as it develops. By which time the damage is done.

But how? The next chapters will give us the answer.

- 13 -

DELAY AGEING PLAN

Happy in mind, healthy in body

Why do more people die of heart attacks on a Monday than any other day of the week?

Or why is 6th January the peak day in the year for calls to the Samaritans helpline?

The answer is an increase in stress hormones – related to anxiety about going back to work. The release of those stress hormones was initially triggered by thoughts.

While most people have no problem in understanding that thoughts and emotions can affect your health, it has not been so easy for biologists to accept[180]. Because, until fairly recently, the way emotions affect health at the cellular level was not well understood.

Every chapter, so far, has had a guiding principle that scientists follow: Show me the evidence and then explain how this may be true. So, when we say that happy people generally stay healthier longer and even live longer, are we straying from 'hard' science into 'soft' science?

I would argue that, without mental wellbeing, the other findings and recommendations in this book would be largely pointless. Surely the aim of extending the years in which you stay healthy is to ensure that they are happy years?

Let's start with some 'hard' science.

Positive people live 7.5 years longer

Becca Levy is Yale Professor of Epidemiology and a leading researcher in the psychology of ageing. Having followed a cohort of 660 older people for 23 years, she and her team published a paper in 2002.

It showed that those who had a positive view of ageing lived 7.5 years longer than those who had a negative view[181].

They followed this up in 2015 with a further study. This showed that people with a negative opinion of growing older had the greatest decline in an area of the brain called the hippocampus which is associated with learning, emotions and memory.

Even more significantly, post-mortems showed that the 'doomsters and gloomsters' had a far greater build-up in the amyloid plaques and tau tangles that signal Alzheimer's than the optimists.

Eric Kim is an age researcher at the Harvard School of Medicine and his website information there states that he *"aims to understand the influence that the social environment has on the connection between psychological well-being and physical health"*. Tracking 7,000 people over the age of 65 for over seven years, he identifies that having a purpose in life is the key factor in healthy mental ageing.

People with purpose and aims take more care of their health and engage better in preventative medical screenings, protective diets and a more active lifestyle. They want to stay around long enough to achieve their goals.

Kim drew some of his inspiration from the Japanese island of Okinawa, a Blue Zone location. There people use the word 'ikigai' which translates as 'the reason you get up in the morning'.

Kim's team recommends that retired people consider volunteer work. That would multiply the economic benefit to society of the healthy ageing strategies we have been exploring. People who delay ageing postpone illness, which costs health services less. But if they then contribute positively, it is a double benefit and a virtuous circle.

But how can your mindset affect your body? What are the mechanisms that take thoughts and emotions and convert them into physical health results?

The brain/body information network

As long ago as 1997, a researcher at the National Institutes for Health called Candace Pert wrote a book called 'Molecules of Emotion'. It caused a lot of excitement and some controversy, because in it she showed how brain and body systems, which had been thought to have little connection, are linked together in a giant 'communication network'.

She demonstrated that the cells in the immune system, the endocrine system (which distributes hormones) and the brain all have similar receptors on them.

These receptors react to neuropeptides which circulate throughout the body and brain. It is these neuropeptides that are the coordinating messengers between the brain, the body and the immune system. This explains how the mind creates physical changes in the brain and body.

As Candace Pert puts it:

"As your feelings change, this mixture of peptides travels throughout your body and your brain. And they're literally changing the chemistry of every cell in your body."

Two personal experiences – one negative, one positive

My own interest in researching health goes back to my mother's sudden death over 50 years ago. It was totally unexpected, and within just months, my father had died of inoperable stomach cancer. I thought, even then, that the emotion of grief had destroyed his immune system and I believe we now know how that happens.

However, the mind can powerfully affect the body in positive ways, too. Back in the 70s and 80s, I was heavily involved in the study of how people learn best and was researching for a book that became 'Accelerated Learning'.

I remember an educational conference in 1981, where an American pentathlete called Marilyn King was speaking. She had been successful enough to be in contention for the 1980 US Olympic team, but a car accident had put her in hospital, with only months to go before the Olympic trials.

Only able to lie in bed, she told how she visualised over and over again the swimming, running, fencing and equestrian moves she would need to make as a pentathlete.

When she was eventually discharged, just three weeks before the trials, her trainers assumed she would have lost coordination and very significant muscle mass due to her extensive hospital stay. She would be out of contention.

Yet they were astonished to discover she had lost so little fitness that she was not only able to compete in the trials, but she was placed second! Unfortunately, she never got to become an Olympian because America boycotted the games in Moscow.

The visualisation and mental imagery had had a direct and measurable effect on her body.

A 2006 study by Vicki Brower showed that mental imagery can translate into physical change via epigenetic change. It concluded:

> *"There is no real division between mind and body because of networks of communication that exist between the brain and neurological, endocrine and immune systems."*

Marilyn King's example is rather extreme, but if you doubt your ability to visualise strongly enough to affect your body, just ask yourself: Have you ever worried?

For example, your son or daughter was late home and you worried. Worry is visualising a possible, but imaginary, negative event, but it is enough to trigger a real physical response in your body. Waking up scared after a nightmare is the same process. The dream events were not real, but your body's reaction is the same as if they were.

Our thoughts, feelings, beliefs, and attitudes can positively or negatively affect our biological functioning. Our minds affect how healthy our bodies are.

Accentuate the positive; eliminate the negative

The modern world makes continuous demands. The always-on "attention economy" keeps us hooked into endless streams of information. It is essential to find moments of stillness, to unwind, to de-stress and touch peace.

Stress accelerates ageing. Studies show that chronic stress (and depression) raise cortisol levels and increase the rate at which telomeres shorten[182][183].

Reduce stress with deep breathing

The following simple 6-step mind calming process helps you get control over your body's autonomous nervous system, reduce anxiety and deal with stress.

A MIND-CALMING MINUTE

1. Stop what you are doing, sit up straight, look upwards, and take in a long deep breath.

2. You probably already feel a bit more relaxed and you almost certainly wanted to smile. The word 'inspire' comes from the Latin *inspirare* – breathe in.

3. Now hold that straight-back position and put one hand on your stomach and the other on your chest.

4. Start breathing in more deeply through your nose, so that you can feel your stomach push out on your hand with every breath in, and go back when you breathe out.

5. Continue to take slow, deep breaths, counting slowly to five. The breaths should push your stomach out first and then lift your chest, filling your lungs full – which our normal shallow breathing never does. See if you can extend the exhale until it is twice the length of the inhale.

6. To release maximum tension, whisper or sub-vocalise the word "c-a-l-m" with every exhale. Or simply elongating the sound of the letter 'M' as you breathe out – a technique charmingly called bumble bee breathing.

As few as 8 rounds of deep breaths will help you reduce stress and anxiety. Breathing this way helps quieten whirling thoughts and brings you back to your body in the present moment.

If you have a challenging task, you can couple this exercise with visualising the task ahead of you. The combination will strengthen your psychological and physical readiness for the challenge. We live in a stressful world and this simple technique is a well-researched stress-buster.

Why does deep breathing like this work? When you feel calm and safe, your breathing naturally slows down and deepens. By mirroring calm, relaxed breathing, you help bring about that same safe state.

You are stimulating the vagus nerve which is part of your parasympathetic nervous system – sometimes called the 'rest and digest system'. This system is complementary to the sympathetic nervous system which includes the fight or flight response.

People have known of the benefits of deep, relaxed breathing for centuries – and have controlled breathing to improve their well-being. The Chinese refer to *qi* (pronounced chi), the Hindus to *prana* and the Greeks to *pneuma*. The Latin word *spiritus* is at the root of both 'spirit' and 'respiration'.

Go outdoors

The benefits of getting outdoors, ideally in a green space, probably do not need reiterating. But if you need a few extra reasons, consider:

- **Phytoncides** are produced by plants. They have antibacterial and antifungal qualities which help the plants fight disease. Scientists think that breathing in phytoncides during a woodland walk – what the Japanese call 'forest bathing' – increases our levels of Natural Killer cells, the white blood cells that help fight off infections and diseases.

- You really should stop and smell the flowers. Research shows that **natural scents** like pine, roses and freshly cut grass make you feel calmer and more relaxed. Free aromatherapy!

- Shorter days and lower light levels in winter can trigger Seasonal Affective Disorder, or SAD – a condition that is marked by symptoms of anxiety, tiredness, and sadness. Being outdoors **increases light levels** and reduces SAD, as do B-complex vitamins.

- **Vitamin D** is essential for a well-functioning body. We get more than 90 percent of our vitamin D from casual exposure to sunlight.

In 2010, researchers at the University of Essex reported results from a meta-analysis that showed just five minutes of 'green' exercise resulted in **improvements in self-esteem** and mood[184].

Get physical

Throughout this book, the critical importance of activity/exercise has already been a consistent theme. That includes walking at a good pace for at least 30 minutes (it can be split into 3 sessions of 10 minutes each) on 5 days a week. And the simple core strengthening exercises we detail in Chapter 14.

There are numerous studies that also confirm the benefits of Pilates, yoga, tai chi or qigong. But not if you skip the class!

Mindfulness

There has been an explosion of interest in mindfulness over the last few years.

Mindfulness has been shown to help relieve stress, help treat heart disease, lower blood pressure, reduce chronic pain, improve sleep, improve satisfaction with life and even alleviate some gastrointestinal issues[185].

We have already seen several studies in this book – including the Dean Ornish research – that have included mindfulness and stress reduction in their protocols. One key element already detailed is deep rhythmic breathing.

Jon Kabat-Zinn is founder and former director of the Stress Reduction Clinic at the University of Massachusetts Medical Center. He is one of the key scientists who have helped to bring the practice of mindfulness meditation into mainstream medicine.

The aim of mindfulness is to achieve a state of alert, focused relaxation by deliberately paying attention to your thoughts and sensations, without judgement. The mind is focused on the present moment – a state of meditation.

There are many excellent websites that will teach you mindfulness, but the basics are simple.

SIMPLE MINDFULNESS

Sit on a straight-backed chair or cross-legged on the floor.

Sit quietly and focus on your natural breathing. Notice the sensation of air flowing in and out of your nostrils. And your belly rising and falling as you inhale and exhale.

Allow thoughts and distractions to come and go – without judgement – and keep returning your focus on breathing.

You can softly vocalise the word C-A-L-M.

Notice any body sensations such as an itch, again without judgement, and let them pass.

Notice the feeling in each part of your body, one by one in succession from your toes, slowly up to the top of your head. Acknowledge sensations and let them go.

Notice and allow emotions to be present, again without judgement. Label any emotion as what it is, be it anger, frustration, happiness or sadness.

Accept the presence of your emotions without judgement and let them go.

If your mind wanders, notice where it has gone and gently redirect it to sensations in the present.

The goal is not to ignore or get rid of thoughts in order to have a 'blank' mind, but simply to <u>notice</u>, with full attention, what arises.

Mindfulness can be especially helpful in situations where outside stressors, over which you have limited control, are creating unhappiness.

I am certainly not an expert, but I can confirm that a mindfulness session of even 15 minutes creates a very calming, peaceful and relaxed feeling. It is a much-needed antidote to the stresses of modern life.

The recommendation is to give yourself the mind calming benefits of a 15-minute mindfulness session on at least one day in the week.

There are links to guided mindfulness/meditation sessions on <u>acceleratedlearning.com/delay-ageing</u>.

But can mindfulness slow ageing?

We've seen how Nobel prize winner Elizabeth Blackburn has linked chronic stress exposure and depression with shorter telomere length – and therefore shorter life[186].

In an article for the New York Academy of Sciences, she and other researchers conclude that:

> *"Meditation practices may promote ... cell longevity both through decreasing stress hormones and oxidative stress and increasing hormones that may protect the telomere".*

I think the answer is yes.

Personal growth and goals

"The purpose of life is a life of purpose" is an aphorism attributed to author Robert Byrne.

The phrase echoes conclusions drawn by the researchers behind the Blue Zones. Yes, their diet is healthy, and daily exercise is built into their life, and they do live in close-knit communities, but they also have a sense of purpose, well into very old age.

So, what meaningful goals have you set recently?

We must each find our own reason to get up in the morning, but it is a key factor in living a long and healthy life. When you allow age to become your identity, you have given up the power to determine your future.

Here are two final thoughts for this chapter:

- Health is not just the absence of disease – it is a state of physical and mental wellbeing.

- Old age doesn't start until you start looking back, rather than forward!

- 14 -

DELAY AGEING PLAN

Keep active

I have some good news for the people whose favourite activity is 'jumping to conclusions'. There are activities that deliver the most benefit for the least time spent.

Since I have a busy and desk-based writing life, staying fit in the least amount of time and for the least effort has been an objective. Not because I am lazy, but because spending an hour a day on activity or exercise is not easy. Hence my emphasis on how much you can achieve in just 37 minutes a day. Of course, that doesn't have to be the maximum time you spend!

Increase heart health and lung capacity

Three times 10 minutes is better than 10,000 steps.

Have you ever thought through the implications of the advice to 'take 10,000 steps a day'? It's about a 5-mile walk, every day!

Well, you will be pleased to know that 10,000 steps never originated from a scientific study. It wormed its way into public consciousness because a Japanese doctor invented a pedometer and named the device *manpo-kei*. This literally translates as '10,000-step meter'. It was an advertising slogan.

Fortunately, we do not necessarily have to walk 5 miles every day to get fit. I am indebted to the UK's TV doctor, Michael Mosley, for a far less challenging fitness routine.

Three 10-minute brisk walks a day

Dr Mosley teamed up with Professor Copeland of Sheffield Hallam University to test two groups of people. One group walked 10,000 steps a day and the other did three 10-minutes brisk walks a day. 'Brisk' meant fast enough to be able to talk, but not sing[187].

The result? The 3 x 10-minute walkers not only found it practical and easy to stick to, but they actually spent more time getting slightly out of breath and increasing their heart rate than the 10,000-step group.

The elevated heart rate, the researchers noted, was the key to the health benefits. Even though they 'only' did about 3,000 steps, they achieved more aerobic value.

In fact, very high intensity exercise may even be counter-productive for the average person. Elite athletes are known to be more vulnerable to upper respiratory tract infections due to compromised immune function after prolonged exercise[188][189].

There is even a theory that the faster an organism uses up oxygen, the shorter it lives, because fast intake of oxygen is likely to increase oxidative or free radical damage. Which would imply that moderate exercise may be better than intense exercise as you get older.

Or running once a week

If you want to up the activity level a bit, in 2019 the *British Journal of Sports Medicine* published an analysis of 14 previously published studies, involving 232,149 runners. They found that even those who reported running as infrequently as once a month had a 27% reduced risk of death from any cause, compared with non-runners.

Elaine Murtagh, an exercise physiologist at Mary Immaculate College in Limerick, Ireland, commented:

> *"This is good news for the many adults who find it hard to find time for exercise. Any amount of running is better than none."*

So, if you can, a weekly or even bi-weekly jog helps you keep fit. Meantime, three 10-minute brisk walks should be a daily target. But walking or running are not quite enough on their own, because to stay fit, you need to add some 'strength training'.

Increase core strength

I realise that the term 'strength training' may be off-putting for some. But all it means is maintaining enough bodily strength to ward off bone loss, joint pain, and loss of balance.

Think toning. Besides, who wants to be weak?

Without some fitness exercise, the average person over the age of 60 can lose as much as 3% of their muscle mass each year. That's 30% muscle loss over a decade! Core strength training and 3 x 10 walking combine to counteract that loss.

Nor does strength training mean a lycra-clad visit to a gym. The exercises we illustrate require no equipment and they support balance, flexibility and mobility, strengthening a range of muscle groups. The priority is to strengthen legs, arms and especially core.

People who lose strength in their **legs** have difficulty getting around and have more difficulty maintaining a healthy weight, because lean body mass – muscle – automatically burns more calories than fat. So, increasing your lean body mass means it is easier to maintain weight.

People who lose strength in their **arms** will eventually have difficulty getting up from chairs or the floor as they get much older.

People who lose strength in their **core** – the muscles in the trunk and pelvis – are susceptible to lower back pain and poorer balance. A strong, flexible core supports almost every aspect of fitness. Core activities that get your heart beating faster and you breathing harder boost your blood flow – and better blood flow and core strength also lead to a better sex life.

Our Fab Four Exercises

This is not a book on fitness, it is a book on extending your healthy years, and there are some very good exercise apps available. But there are four basic exercises that almost anyone can do, and which will keep them fit.

Since I don't know your age or level of fitness, I am not suggesting you do them without supervision or an OK from your doctor.

However, they are likely to be included in any local or online Pilates class repertoire. The advantage of being in a class is that the leader will supervise you and ensure you don't try too much, too soon — and you'll have the group dynamic to motivate you. Other very good alternatives can be yoga, tai chi or qi gong.

Before you start any exercise, remember the vital importance of warming up and cooling down, and if you have osteoporosis, back problems, or any other health issues, talk to your doctor before doing any exercise.

I can do all four of these exercises in about 7 minutes. I have included them to illustrate the fact that including core strength exercises into your day need not demand a lot of time.

THE FAB FOUR EXERCISES
THE PLANK

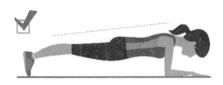

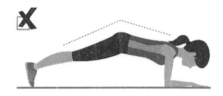

Rest your forearms on the floor, with your elbows directly underneath your shoulders and hands facing forward so that your arms are parallel.

Extend your legs out behind you and rest your toes on the floor. Your body should form one straight line from your shoulders to your heels. Don't raise your bottom.

Squeeze your entire core to keep your lower back straight. Make sure you are not dropping your hips or pushing your bottom up high toward the ceiling.

Position your head so that your neck is in a neutral position and your gaze is on your hands.

Hold this position for at least a minute, ideally 90 seconds.

THE FAB FOUR EXERCISES
THE BRIDGE

Lie on your back with your hands at your sides, knees bent, and feet flat on the floor under your knees.

Tighten your abdominal and buttock muscles.

Raise your hips so that your knees and shoulders are in a straight line

Squeeze your core and pull your tummy button back toward your spine.

Hold for 20 to 30 seconds, and then return to your starting position.

Complete at least 5 repetitions.

THE FAB FOUR EXERCISES

THE CRUNCH

Lie down on the floor on your back and bend your knees, placing your hands across your chest. It is VERY important not to put strain on your neck.

Pull your belly button towards your spine.

Slowly contract your abdominal muscles, bringing your shoulder blades slightly off the floor – about 1 or 2 inches, 3 to 5 cm.

Breathe out as you come up and keep your neck straight. Imagine you're holding a tennis ball under your chin. That's the angle you want to keep your chin at during the exercise.

Hold your position at the top of the movement for a few seconds, breathing continuously.

Slowly lower back down, but don't relax all the way.

Complete 10 repetitions.

THE FAB FOUR EXERCISES
THE LUNGE

Stand tall with your legs slightly apart and spine straight. Look directly ahead.

Step forward with your left leg and bend both knees at a 90-degree angle. Make sure your right knee does not touch the ground. The toes of your right foot will be on the ground and your heel faces backwards.

Your body should still be erect without leaning over.

Stay in this position for 10 seconds and then go back to the start point. Now, bring your right leg forward and repeat the process. Feel the stretch in your heels, your 'quads' (on the front of your thighs) and 'glutes' (the muscles that shape your bottom).

Keep any pressure on your knees to a minimum. If you feel the strain too much, take shorter lunge strides.

Some people add dumb-bells and curl their arms up and down to work their upper body at the same time.

The four exercises above represent a small range that most senior people can do, even if you may have to modify them a little depending on your starting level of fitness.

If you are looking for a free phone app that shows you how to do a range of fast, effective and equipment-free routines and builds a programme for you based on your personal goal, check out *7-Minute Workout*. There are illustrations to ensure you get each exercise right and avoid injury.

The surprising benefits of standing on one leg

You need strong muscles and core strength for good balance. But balance tends to worsen as we age, as our eyesight deteriorates, the sensory input from joints and muscles is less reliable, and we lose cells in the inner ear that detect movement.

One in three over-65s will have at least one fall a year, according to the NHS – and it's 50% in the over-80s. Although low blood pressure and heart disease are causes, poor balance is an important reason. Weak ankles are a common cause of trips and falls, because if you go over on your ankle, the damaged ligaments may heal in a way that leaves them slacker and more prone to further injuries and falls.

Improving your balance and ankle strength can help prevent those falls and the fractures, especially hip fractures, that may result. A survey by Age UK found that 36% of older people named fear of a fall as their major health concern.

It may sound a little odd, but standing on one leg is a simple, but very effective exercise for improving balance. Experts suggest you can do this by standing on one leg as you brush your teeth, swapping feet half-way through.

A brain health check from one-leg balancing

Kyoto University researchers enrolled almost 1,300 men and women aged about 67 and asked them to stand on one leg, with their eyes open, and maintain their balance for about 20 seconds. They were then tested for brain health. The results are disconcerting[190].

More than 30% of the subjects who had trouble balancing themselves for this length of time were found to have either 'cerebral small vessel disease', or tiny haemorrhages, or both.

Haemorrhages in the brain can lead to strokes. Cerebral small vessel disease can lead to a condition where inadequate blood supply causes a tissue to die and is linked to cognitive decline and the development of dementia, Alzheimer's disease, and Parkinson's disease.

Dr Yasuharu Tabara, who led the research at Kyoto, says:

> *"The ability to balance on one leg is an important test for brain health."*

The UK's Medical Research Council found that 53-year-olds who could stand on one leg for ten seconds, with their eyes closed, were the most likely to be fit and well in 13 years' time[191].

How good is your balance?

A study published in the *Journal of Geriatric Physical Therapy* measured how long different age groups could averagely balance on one leg to establish what could be considered 'normal'[192]. Delightfully, they refer to it as the 'uni-pedal stance test'.

To do the test, time how long you can stand on one leg, with your hands resting on your hips. Stop the clock, or your counting, when your raised foot touches the floor or your other leg, or you have to lift your arms off your hips to steady yourself.

This is what is considered 'normal' for your age.

Age	Eyes open	Eyes closed
Under 40	45 seconds	15 seconds
40-49	42 seconds	13 seconds
50-59	41 seconds	8 seconds
60-69	32 seconds	4 seconds
70-79	22 seconds	3 seconds
80-99	9 seconds	2 seconds

But resist being smug at your result. As a publicity stunt, contestants in China competed to win a new car, the winner being the one who could stand on one leg for the longest while touching the car. The keys were won by a man who managed it for 7 hours.

Even that feat pales, compared to the multiple Guinness World Record entrant, Suresh Joachim. He balanced on one foot for 76 hours and 40 minutes.

Activity is a key part of postponing ageing

Inactivity is a risk factor for heart disease, cancer and diabetes. A 2018 study showed that frequent TV-watching for long periods was linked to a significant increase in coronary heart disease[193]. Another study claims that every hour of TV-watching can cut 22 minutes off your life![194]

Of course, we are all going to watch TV, but getting up every hour and walking around for a couple of minutes is a sensible antidote.

If exercise were a drug, we would all be taking it. Because those daily 37 minutes will increase autophagy, reduce your risk of heart disease and cancer, increase brain health, strengthen your lungs, regulate your blood pressure, improve your mood, strengthen your bones and muscle, help control your weight, reduce blood sugar levels, improve skin tone and improve your sex life.

Sign me up!

- 15 -

DELAY AGEING PLAN

Fewer calories = Longer healthier life

Almost all longevity researchers agree that '**calorie restriction**' increases lifespan – at least in laboratory animals.

The research has been ongoing for decades, and has been conducted on a variety of animals, mainly worms, crabs, snails, fruit flies and rodents.

In these studies, when test animals were given 10% to 30% fewer calories than usual, <u>but provided with all necessary nutrients</u>, many showed a significant extension of lifespan and reduced rates of several diseases, especially cancers.

That's not surprising. We have already seen that one of the ways to counteract the problem of senescent cells, DNA damage, the loss of stem cells, weaker nutrient sensing and accumulated protein errors is calorie restriction.

But our objective is not increased lifespan at any cost, but rather an increased <u>healthy</u> lifespan. And unless the plan to achieve it is practical and enjoyable, few will follow it.

I believe the phrases 'calorie restriction' and 'intermittent fasting' are both negative and unhelpful – because there are sensible ways to reduce calorie intake that do not involve sacrifice and are easy and effective.

Nevertheless, Professor Luigi Fontana of Sydney University cites excess calorie intake as <u>the</u> root of ageing in his new book 'The Path to Longevity'.

Excess calories are the root of unhealthy ageing

Your metabolism slows down as you get older. This is best and rather dramatically illustrated with a graph.

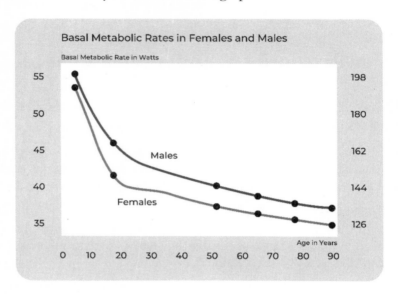

It shows the average basal metabolic rate at different ages. This is the number of calories you need when at rest, just to keep your body processes going – breathing, producing proteins, hormones and cells, and metabolising food. On top of that, you will be using up calories through activity.

The problem is that most of us, as we get older, tend to move less, so the calories we burn in activity goes down, at the same time as basal metabolism also reduces.

If we eat the same amount of food as before, the result must be that we take in more calories than we need – and increase our weight. The trouble is we do not, mostly, eat less to compensate.

The reason is deeply embedded in evolution. Throughout all the thousands of past years of human history, the problem was getting enough food. So, we are not only programmed to eat when we are hungry, we are programmed to continue to eat beyond enough: 'just in case the next meal isn't going to be easily available'.

We are also programmed to like sweet and fatty foods because they are the fastest source of energy which, until the very recent past, was what humans needed.

Add the fact that most jobs today have minimal physical content and that most transport no longer involves walking, and you can see why Professor Fontana identifies excess calories as a problem.

Excess calories – which is another way of saying excess food intake – are ageing in at least three important ways.

First, that extra food needs to be metabolised, which leads inevitably to more free radical production as food is burned with oxygen. Free radicals damage mitochondria and, as we have seen, damage to mitochondria is a hallmark of ageing.

Second, excess calories increase blood sugar (glucose), which raises insulin levels and, over time, decreases insulin sensitivity and increases inflammation. Decreased insulin sensitivity (and inflammation) are more hallmarks of ageing, and provide a fertile environment for cancer to spread.

Third, overeating can lead to excess body fat – becoming overweight or obese. Fat cells increase in size and produce hormones called adipokines, which trigger inflammation and insulin resistance. This is a risk factor for cancer, heart disease, stroke and dementia.

Positive health outcomes of calorie reduction

In contrast, a reduction in calories:

1. Lowers glucose levels in your system. Low glucose levels are strongly linked to healthy ageing.

2. Helps activate autophagy to clear senescent cells.

3. Challenges the body to go into a conservation mode that turns on cell and DNA repair processes.

4. Helps keeps your weight under control.

5. Helps reduce fat in the liver, which is a risk factor for type 2 diabetes.

6. Reduces blood sugar levels, which increases the availability of NAD+ and therefore sirtuin activity.

7. Should help you sleep better.

8. May suppress brain inflammation (from animal studies).

9. Improves your sensitivity to insulin and IGF-1 (the hormone insulin-like growth factor).

The lowering of insulin (and its sister hormone IGF-1) is extremely important because this indirectly turns on important 'longevity genes' called FOXO genes. That, in turn, triggers many other processes that delay ageing.

This sequence summarises them:

Lower calories → Reduces insulin →
Reduces mTOR → Turns on FOXO genes →
Increases autophagy and DNA repair
+ Reduces loss of stem cells
+ Improves the ability of T-cells and Natural
Killer cells to kill cancer cells

So, fewer calories help delay ageing – but how can you achieve that without being hungry?

Say 'Hara Hachi Bu'

You will remember that Okinawa is one of the Blue Zones, where people are fit and healthy well into their 90s and even 100s.

Analysis of their diets reveals that they typically eat 10% less than equivalent-height Westerners. Their diet features nutrient-dense, fibre-rich and calorie-light plant foods, and they also practise 'Hara Hachi Bu' – which means 'eating until you are 80% full'.

That automatically decreases calorie intake.

But how would you know when you are 80% full? Dan Buettner (the author of *Blue Zones*) explains it as stopping eating when you can say: "I am not hungry any more", rather than stopping eating when you say: "I am full."

Here's how you can adopt Hara Hachi Bu into your daily routine – and even improve on it.

Eat slowly and really concentrate on your food. If you take time to *really* savour the taste and texture of your food, you will easily sense the difference between being full (100%) and no longer being hungry (80%).

Slowing down eating is key because it takes about 10 minutes for hormones, secreted as you digest the food, to register a sense of satiety or satisfaction in your brain.

Don't eat and watch TV, or eat and read. The 'satiety signal' that you are no longer hungry will not register until too late.

Put your knife and fork down occasionally and consciously sense how satisfied you feel.

Eat from smaller plates, which makes it even easier. This simply changes your perception of what is a filling meal.

Emphasise protein and fibre foods to reduce net calorie intake. On average, a person uses about 10% of their daily energy expenditure in digesting and absorbing food. But this percentage is very different depending on the type of food you eat[195][196].

Protein takes the most energy to digest – some 20 to 30% of total calories in eaten protein go to digesting it.

Next are carbohydrates (5 to 10%) and then fats (0 to 3%). Almost all calories from sugar and sugary drinks are available immediately.

This means that if you consume 100 calories from protein, your body uses about 25 of those calories to metabolise the protein. So, you are left with a net 75 calories. The same 100 calories of pure carbohydrate would leave you with a net average of 93 available calories, and pure fat would give you a net available 97 calories.

Fibre does not have any available calories.

So, by increasing the proportion of protein (modestly) and fibre (substantially) in your diet, you automatically reduce your net calorie intake. Better still is to increase protein from plant foods. The foods listed in the **Delay Ageing Food Plan** are chosen to help you do all this – they are nutrient dense, high in fibre and net calorie light.

When you eat matters – daytime is better. Your body clock or circadian rhythm regulates your sleep patterns, hormone levels, body temperature – and metabolism.

Because we are daytime animals, our metabolism is higher in the daytime than at night.

Several studies have indicated that eating earlier in the day, and only having a light last meal at dinnertime, is an easy way to reduce net calories (and control your weight).

One particularly striking study allowed two groups of overweight women the same number of calories a day, but half the cohort consumed 50% of those calories at breakfast. In the other half, 50% of the calories were eaten at dinner.

The result? The breakfast eaters achieved over twice the amount of weight loss compared to the bigger dinner eaters. At the end of the 12-week study, they also recorded significant reductions in triglyceride (blood fat) levels, cholesterol levels and insulin levels [197].

Another smaller study[198] found that energy levels were higher for early eaters and sleep was improved.

Of course, this pattern – eat like a king at breakfast, lunch like a prince and take dinner like a pauper – is common to many cultures, and Blue Zone dwellers. However, it is almost the opposite of how most Westerners eat, due to work lifestyles.

Allow at least a 12-hour gap between dinner and breakfast. This helps your system utilise all its available glucose. Once this happens, your body switches to using fat as an energy store. Studies indicate a gap of 15 -16 hours could be even more effective[199].

If you combine Hara Hachi Bu with the other recommendations above, then I believe you are creating the precisely the conditions of reduced calorie intake that researchers into healthy ageing are seeking, but without the stress and bother of a calorie-counting diet, or invoking the unappealing word 'fasting'.

We have mentioned before that the diabetes drug metformin is being studied by several age researchers. The primary effect of metformin is to control – lower – blood sugar levels. I believe most people could achieve a similar result from enhanced Hara Hachi Bu.

Evidence for positive effects of calorie reduction

Rhesus monkeys trial 2009

In 2009, the journal *Science* published the results of a 20-year study where 38 rhesus monkeys had been put on a reduced calorie diet and were fed a nutritional supplement to ensure no nutritional deficiency. Their results were compared to a control panel of 38 monkeys allowed to eat as much they wished.

The monkeys, of course, are a much closer model for humans than the usual subjects used in longevity studies – like fruit flies, nematode worms and rodents.

The results were impressive. The test monkeys not only looked younger, but were assessed as biologically younger than the control monkeys. They lived about 25% longer – many to the equivalent of 120 in human years.

More importantly, they had far fewer illnesses right up to their deaths – less cancer, less heart disease and no type 2 diabetes. They retained both body muscle mass and brain grey matter and performed better on cognitive tests[200].

But can we expect similar results in humans?

CALERIE trial 2018

The US National Institute on Aging (NIA) conducted a clinical trial called 'Comprehensive Assessment of Long-term Effects of Reducing Intake of Energy (CALERIE)'. It reported in 2018[201].

218 adults were randomly divided into two groups. People in the experimental group were told to follow a calorie-restriction diet for 2 years, while those in the control group followed their usual diet.

Participants in the experimental group reduced their daily caloric intake by 12 percent – which is about what you would achieve if you followed the advice above.

Compared to participants in the control group, those in the lower calorie group had reduced risk factors (lower blood pressure and lower cholesterol) for age-related diseases such as diabetes, heart disease, and stroke.

They also showed decreases in inflammatory markers and thyroid hormones which are associated with longer lifespan and diminished risk for age-related diseases.

The report concluded:

> *"In the calorie-restricted individuals, no adverse effects (and some favorable ones) were found on quality of life, mood, sexual function, and sleep".*

Although the NIA study on calorie reduction did cause slight declines in bone density, other studies have found that combining physical activity with calorie restriction protects against losses of bone and muscle mass.

For those readers who are interested in actually losing some weight, you should note that the NIA study showed that participants who needed to slim lost on average 10 percent of their body weight over 2 years.

To put this into context, a 10% reduction in body weight is equivalent to a reduction of almost 17 lb or 8 kg for an overweight woman (starting at 170 lb/12 st 2 lb/77 kg) and 20 lb or 9 kg for an overweight man (starting at 200 lb/14 st 4 lb/91 kg).

A follow-up study two years after the intervention ended found that participants had sustained much of this weight loss.

Enhanced Hara Hachi Bu – the ideal diet?

This suggests that Enhanced Hara Hachi Bu might just be the simplest and easiest ever way to maintain a healthy weight.

In a study on intermittent fasting, a group was put on a 900-calorie diet for five consecutive days a month for 12 months. They succeeded in lowering inflammation and blood pressure, and in increasing the number of circulating stem cells. On average, people who normally ate 2,000 calories were reducing their intake by about 1,100 calories a day for those five days, or a total of 5,500 calories in the month.

Enhanced Hara Hachi Bu would result in a continuous 10% reduction in calories a day or about 200 calories a day. Over a month, this is a slightly higher (6,000) calorie reduction than in the Intermittent Fasting study.

Of course, it is possible that a 5-day, much deeper, calorie restriction may produce better results than a 30-day slight calorie reduction. However, I suggest the easier tolerability of Enhanced Hara Hachi Bu will make it a more practical solution for most people.

The Okinawa evidence

We have evidence from Okinawa that calorie restriction does help improve healthy lifespan. Per 100,000 people, there are 54 centenarians on that island, versus 20 per 100,000 in the UK, and 17 in the US.

On Okinawa, among women, mortality from breast cancer is five times lower, and among men, death from prostate cancer is seven times lower than in Westerners. Of course, it is not just their lower calorie intake that accounts for their extended health-span, but also their diet which is low in animal foods, high in vegetables.

Our personal experience

On a personal note, my wife (age 62) and I (80) have followed the advice in this chapter and the food plan for over 18 years.

The food plan is a very-high-nutrient-density diet because, on top of Blue Zone type foods, it also includes a supplement with extra vitamins, minerals, carotenoids, polyphenols and omega-3. During that time, our weight has never altered from a healthy BMI of 24.

Since this is a sample of two, it obviously has no scientific significance. But I hypothesise that we rarely feel the urge to eat more than we need*, partly because our brains have never sensed any nutritional deficiencies.

*other than at parties

- 16 -

DELAY AGEING PLAN

Power nutrients

Even vaguely health-conscious individuals are aware of the 5-a-day campaigns in both the UK and USA, which recommend eating at least 5 portions of fruits and vegetables a day.

Originating in 1988 in California, the 5-a-day message was adopted by the UK Department of Health in 2003. It has been a moderate success in that the current average is almost 3.6 portions a day[202].

The problem is that even 5-a-day does not meet the level or range of vitamins and minerals, and especially flavonoids and polyphenols, that will truly delay ageing.

5-a-day should be 10-a-day

The American Cancer Society[203] and Imperial College London[204] have independently calculated the optimum level of fruit and vegetable intake that makes a real difference to health.

They both agree that it is between 9 and 10 portions of fruits and vegetables a day.

In addition, other institutions recommend 2 portions of omega-3 rich oily fish a week[205].

At those levels, your blood chemistry becomes highly protective. Inflammation declines, DNA and mitochondria repair improve, positive gene expression occurs, autophagy clears senescent cells, insulin sensitivity is increased, sirtuin genes are turned on and protein errors decline.

It is also about the level found in the Blue Zones.

If you regularly eat 9-10 portions of fruits and vegetables a day (and the equivalent of at least two portions of oily fish a week), you don't need a nutritional supplement.

Unfortunately, 9-10 portions of fruits and vegetables a day, or 70 portions a week, is a real challenge for even most the most conscientious of us. Which leaves a gap.

Filling the gap – a simple multivitamin will not do

A simple vitamin and mineral pill will not fill that gap, which is why the evidence is weak that taking a typical A-Z supplement at 100% of RDA/NRI (recommended daily allowance/nutrient reference intake), on its own, can much improve health. (See Chapter 17.)

But what if you could take a food plan that includes all the foods that we have featured, and then top it up with supplements that boost the level and range of vitamins and minerals, <u>and</u> of the most protective plant nutrients, methyl donor nutrients, anti-inflammatories, antioxidants and senolytic nutrients?

Then, I believe even the average 5-a-day person could begin to meet the range and level of nutrition needed to truly delay ageing.

The foundation would still be a healthy diet, because that's always the priority, and supplements do not <u>replace</u> food – they don't provide you with fibre, for example.

But a properly formulated food supplement would really help fill the gap between the realistic and the ideal, which is, after all, what a 'supplement' is for.

The ***Delay Ageing Plan*** does recommend supplementation with certain nutrients. Which nutrients to consider is the subject of this chapter.

Power nutrients from plants

A lot of the most powerful age-delaying nutrients come from pigments in the skins of plants – which they develop as protection from potential damage from the sun and insects. We 'inherit' that protective effect when we eat the plants.

These pigments include **anthocyanins** which give red, purple and blue plants their colour. Anthocyanins have anti-inflammatory and anti-bacterial activity and help to lower blood pressure, reduce cancer cell proliferation and inhibit tumour formation.

They also include **carotenoids** like beta carotene, which gives plants an orange colour, lycopene (red fruits and tomatoes) and lutein (yellow). Other pigments are **betacyanins** (in beets/beetroot) and, of course, **chlorophyll** (green).

Although there are hundreds of **polyphenols** and **flavonoids** (a class of polyphenols), some are much better studied than others and a few appear to have considerably more consistent and significant health benefits.

Below is a summary of plant nutrients that should feature in your diet – including the ones I believe are worth supplementing. I have marked these with a star *.

I have concentrated on polyphenols because we all know that vitamins and minerals are essential for health and there is plenty of information easily available on the benefits of each one. But far less is known about polyphenols, which have a huge impact on your health.

The list reinforces the fact that if you just take a simple multivitamin, you will miss out on a whole category of nutrients that can support your health.

*CURCUMIN/curcuminoids (from turmeric)

Curcumin is a very powerful **antioxidant** and **anti-inflammatory**, which is as effective as many anti-inflammatory drugs[206]. Curcumin also increases the activity of your body's own antioxidant enzymes, especially glutathione[207].

Curcumin blocks a signalling molecule called NF-kB that turns on genes related to inflammation, so it is thought to be **heart protective**[208]. This is likely to be true, as curcumin leads to improvements in endothelial function.

The endothelium is the single-cell lining of various organs including your arteries and blood vessels, and releases substances that regulate blood pressure, and blood clotting[209].

Since inflammation is linked to depression, scientists hypothesised that curcumin could reduce **depression**. A randomised trial in 2014 indicated that curcumin could be equally as effective as Prozac[210].

Depression is linked to low levels of a compound called Brain-Derived Neurotrophic Factor (BDNF) – see Chapter 18.

In animal trials, curcumin boosts the body's ability to create new brain cells by increasing BDNF. So, curcumin should have a role in defending against brain ageing and **Alzheimer's**.

Studies have shown that curcumin can also help prompt the death of cancerous cells and reduce the growth of new blood vessels in tumours and metastasis (the spread of **cancer**).

If all this was not enough, the anti-inflammatory properties of curcumin make it an ideal ingredient to damp down joint pain in **arthritis** supplements[211].

It's hardly surprising, therefore, that curcumin is being researched as a key supplement to slow down ageing[212], not least as it turns down the mTOR and IGF-1 pathways, which, when activated, lead to ageing.

However, curcumin is not well absorbed into the bloodstream. It helps to consume black pepper with it. Black pepper contains piperine, a natural substance that appears to enhance the absorption of curcumin by some 2,000% or 20 times[213]. If taken as a supplement, ensure there is piperine alongside curcumin, and take it with a fat source to further enhance bio-availability. Ideal would be omega-3.

Get curcumin from: Turmeric (but curcumin levels are only about 3% - 5% in turmeric).

---

*GREEN TEA and green tea extract

Green tea has multiple benefits, largely because it contains a powerful natural antioxidant and anti-inflammatory polyphenol called epigallocatechin-3-gallate (EGCG).

A review of several large-scale epidemiological studies shows that women who drink the most green tea appear to have a 20-25% lower risk of **breast cancer**[214]; that green tea drinkers may be 40% less likely to develop **colorectal cancer**; and that men drinking five or more cups of green tea is linked to a lower risk of advanced **prostate cancer**[215]. This may be, at least partly, because EGCG has been shown in lab tests to damp down the proliferation of cancer cells.

Further population studies show an inverse correlation between green tea consumption and dementia, including **Alzheimer's**. A major review in the journal *CNS Neuroscience and Therapeutics* suggests that green tea exerts this neuro-protective effect partly via its ability to counteract free radicals, and partly because it improves cell signalling[216 217].

Green tea also contains a compound called L-theanine which can increase serotonin and dopamine levels. This suggests it will have a **brain calming** effect – which is indeed often reported by users. Furthermore, animal studies link consumption of green tea to improvements in learning and **memory**[218].

A meta-analysis of 7 studies involving over 280,000 people indicates that green tea (and coffee) consumption may lower the risk of **diabetes**[219]. Another meta-analysis in the *American Journal of Clinical Nutrition* suggests this is due to green tea's ability to improve insulin sensitivity and help to *"significantly reduce insulin concentrations"* and blood sugar levels[220].

Since green tea has a proven antioxidant effect, it is not surprising that it has been found to protects LDL cholesterol particles from oxidation, which is one of the pathways to **heart disease**[221].

Indeed, a Japanese study of 40,500 older people over 11 years links consumption of 4-5 cups of green tea a day to a reduction in mortality due to cardiovascular disease, stroke and indeed all causes[222].

Throughout this book, we have seen many ways in which green tea can help counteract the drivers of ageing.

One conclusion from the literature is that the effective level of consumption of green tea is around 3-4 cups a day. Whilst that is common in Japan, most Britons and Americans would be unlikely to reach that level – hence green tea extract is recommended in a comprehensive supplement.

Get EGCG from: green tea or green tea extracts

---

*BETAINE

Betaine works quickly to help prevent the build-up of an amino acid called homocysteine. This amino acid can harm blood vessels and contribute to **heart disease**, **stroke**, and circulation problems.

In animal models, betaine has been shown to reduce beta amyloid levels and **Alzheimer's** symptoms[250]. A review article in the *American Journal of Clinical Nutrition* concluded:

"The growing body of evidence shows that betaine is an important nutrient for the prevention of chronic disease".

Get betaine from: Beetroot, spinach and whole wheat grains.

PRO-ANTHOCYANIDINS like Grapeseed, Bilberry

A high intake of antioxidants, such as pro-anthocyanidins, is generally associated with a reduced risk of various **cancers**.

Pro-anthocyanidins help prevent E. coli bacteria from sticking to the walls of the bladder and urethra. Hence, they help prevent **urinary tract infection**. They may also increase the strength of blood vessels in people with hypertension and **diabetes**[223].

*Grapeseed extract is one of the best sources of pro-anthocyanidins. Grapeseed extract has been shown in a meta-analysis to help blood flow and reduce blood pressure[224]. It also helps reduce the potentially dangerous oxidation of LDL cholesterol [107 225].

Grapeseed extract's anti-inflammatory effect in the brain probably explains another benefit, which is to improve attention and **memory** in older, healthy adults[105].

Finally, in lab tests, grapeseed extract has been shown to inhibit multiple **cancer** cell lines[226].

Get pro-anthocyanidins from: Grapes, grapeseed, cranberries, bilberries, strawberries.

---

*BETA CAROTENE

Beta carotene converts to vitamin A as needed, and so is important to a strong **immune system** and, combined with lutein, to healthy **vision**. It is an antioxidant that fights free radicals. A study of 4,000 men indicates long term supplementation slows cognitive decline, probably related to its antioxidant capabilities.

Get beta carotene from: Carrots, sweet potatoes, kale, spinach, red and yellow peppers, apricots. Ideally, consume with a fat source as beta carotene is fat soluble. One study indicated that smokers should not supplement with beta carotene as it raised their risk of cancer, but several researchers dispute the findings of that study.

---

*LUTEIN and *ZEAXANTHIN

There is convincing evidence that these carotenoids help halt or slow **macular degeneration** and **cataracts**[227].

They are important antioxidants that have been shown in animal studies to help protect against **atherosclerosis** – the build-up of fatty deposits in arteries, a condition that leads to heart attacks and stroke[228]. They also help recycle glutathione[229].

Get lutein and zeaxanthin from: Kale, spinach, chard, broccoli, collard greens, eggs, corn, pumpkin.

---

*LYCOPENE

A powerful antioxidant which can help prevent or slow down the progression of some types of cancer, especially **breast and prostate cancers**, by limiting tumour growth[230 231].

Metabolic syndrome is a dangerous condition, which is a combination of high blood glucose levels, plus high blood pressure plus high blood fat levels. In a 10-year study, patients with **metabolic syndrome,** but who also had above-average lycopene levels, showed reduced mortality[232].

Lycopene also appears, in animal experiments, to reduce **dementia** risk due to its significant antioxidant effect.

Get lycopene from: Tomatoes – but to make lycopene more bio-available, it is better to cook them, ideally in olive oil. Other sources include sundried tomatoes and tomato puree. Guava has some lycopene, as has watermelon, but at levels far below tomatoes.

---

*SOY ISOFLAVONES

These are phytoestrogens ie. plant oestrogens, which have a weak oestrogen-like action. Studies indicate that soy isoflavones confer a benefit for **heart health** and for the maintenance of **bone health** in post-menopausal women[233]. The three main isoflavones are genistein, daidzein and glycitin.

Genistein has important **anti-cancer** properties. A frequently cited review in the journal *Cancer Letters* highlighted genistein's ability to help inhibit cancer cell growth and force cancer cells to self-destruct (a process called apoptosis). It concluded that genistein was a candidate for adjunct cancer therapy[234].

Genistein also shows ability to alleviate hot flushes/flashes during the **menopause.**

A large meta-analysis in 2018 showed that soy isoflavones can help improve **cognition** in older adults, probably as a result of their anti-inflammatory effect[235].

A Harvard School of Public Health summary concludes:

> *"Soy is a nutrient-dense source of protein that can safely be consumed several times a week and is likely to provide health benefits—especially when eaten as an alternative to red and processed meat."*

Get isoflavones from: Soybeans (aka edamame beans when green and young) and their products like tofu, miso, soy sauce, natto, soy mince, soy milk, soy yogurt; chickpeas, peanuts and pistachios.

---

*FUCOIDAN

Brown seaweed is a common staple in many Asian diets, including in Okinawa. It is rich in fibre, minerals and polyphenols. Traditionally used to treat thyroid conditions, because of its iodine content, brown seaweed also contains a compound called fucoidan.

Data from the Sloan Kettering Cancer Centre[167] shows fucoidan can reduce inflammation and improve the immune system. Importantly fucoidan also appears to help mitigate the effect of viral infection where lung damage is involved. This type of damage occurs with Influenza A, which has been responsible for at least three pandemics last century, and coronaviral infections[168].

According to the Sloan Kettering report, fucoidan also may demonstrate anti-tumour effects[236]

Fucoidan is found in several species of brown seaweed, including ascophyllum nodosum (Norwegian kelp), fucus vesiculosus (bladderwrack) and undaria pinnatifida (wakame).

FISETIN

This nutrient helps clear senescent cells and was the top **senolytic** out of 10 tested in a 2018 *Lancet* published study[6].

Alzheimers.net rated strawberries as the number one fruit for brain protection – albeit from animal studies [6 237].

Get fisetin from: Strawberries, apples, mangoes, kiwifruit, peaches.

QUERCETIN

An antioxidant that helps quench free radicals and fights inflammation. May be as effective as resveratrol, but more bioavailable[238]. Quercetin is widely available in many fruits and vegetables and appears to have **anti-cancer** effects in prostate cells[239].

Get quercetin from: Onions, apples, leafy green vegetables, broccoli, green tea, cherries.

ELLAGIC ACID

Ellagic acid works with other nutrients like quercetin, glutathione and vitamin C, to boost the effect of phase 2 enzymes[240].

Phase 2 enzymes are a very important part of your natural mechanisms for resisting **cancer**. They speed up the elimination of toxic compounds that could otherwise trigger cancer. Ellagic acid also encourages cancer cells to self-destruct – a process, as we have seen, called apoptosis.

Get ellagic acid from: Raspberries, strawberries, blackberries, grapes, pomegranate, pecans, walnuts.

CHOLOROGENIC ACID

This nutrient helps lower blood glucose concentrations. May help reduce the oxidation of LDL cholesterol and possibly break down the fatty sludge in arteries that leads to atherosclerosis.

Get chlorogenic acid from: Apples, pears, coffee, tomatoes, blueberries.

APIGENIN

Apigenin is a widely available natural flavonoid shown to have anti-inflammatory and antioxidant properties.

When apigenin was given to obese mice, it prevented the breakdown of NAD+ and therefore increased levels of SIRT1, a protein involved in healthy mitochondrial function. Other papers indicate *"it is a promising molecule for cancer prevention"* [241].

Get apigenin from: Many fruits and vegetables, but it is particularly abundant in chamomile, parsley and celery.

GLUTATHIONE

This is a very important natural antioxidant that the body makes itself, but levels decline with age. So, a role of food and/or supplements is to boost your own ability to increase this 'master antioxidant'. Taking glutathione as a supplement is not effective, although there are a few foods that contain it directly.

Since sulphur is needed to make glutathione, increase sulphur-rich foods. Also increase antioxidant-rich foods and supplements, so that your own internally created glutathione has less to do.

A placebo-controlled study showed that a 500mg vitamin C supplement raised glutathione levels by almost 50% within three weeks[242].

Selenium is a co-factor (ie. it is needed) for glutathione production, so also eat selenium-rich foods, which include grains and seafood, and ensure any supplement includes selenium.

Curcumin also boosts glutathione production[207].

Finally, ensure enough exercise and sleep, as they are major factors in keeping glutathione levels healthy.

Get glutathione from these foods: Spinach, avocados, turmeric.

Boost glutathione production with: Sulphur foods: onions, garlic; Cruciferous vegetables like broccoli, brussels sprouts, cauliflower, kale, watercress; Selenium foods, including seafood, fish, brazil nuts.

SPERMIDINE

This polyamine (meaning it combines two amino acids) appears to mimic the effect of calorie restriction and triggers autophagy[7].

Get spermidine from: Wheatgerm, soybeans, mushrooms (especially shiitake), blue cheese, aged cheddar cheese, peas, some nuts, fermented foods including sauerkraut and miso.

RESVERATROL

Resveratrol features strongly in the literature on life extension. It does extend life in nematode worms and fruit flies, but the research has almost always been done at high doses on cells and animal models.

A study on 268 people in the Chianti region of Italy, who consume a lot of plant foods with resveratrol, found that it did not show up in their blood and there was no clear effect on inflammation markers, cardiovascular health or mortality[243].

This links to the fact that we know it is not easily bioavailable[244], although combining it with a fat source helps. Nevertheless, resveratrol-rich foods are well worth including in your plan, as they contain other powerful phytonutrients, too.

Get resveratrol from: Grape skin and seeds, grapes, blueberries, peanuts, cranberries, dark chocolate, red wine.

FIBRE

Fibre, of course, is not a polyphenol. But good levels of soluble and non-soluble fibre are vital in staying healthy[245]. The average person, however, eats only about half the recommended level which is 25g a day for women and 38g for men.

Fibre is a key influence on the composition of your microbiome. Our recommended food plan is high in fibre, including insoluble fibre. Insoluble fibre encourages the growth of the strains of bacteria which create butyrate, which in turn reduces inflammation and stimulates the production of regulatory T-cells. Immunologists call them 'Tregs'.

Tregs regulate other cells in the immune system and control the immune response to self and foreign particles. So, they help prevent allergies and autoimmune disease[246].

The reduction in fibre in the Western diet has been cited as a contributory factor to the rise in auto-immune health issues like irritable bowel disease, type 1 diabetes, multiple sclerosis, psoriasis and rheumatoid arthritis.

A meta-survey in the *British Medical Journal* also shows that higher fibre intake clearly correlates to a lower cardiovascular disease risk[247].

Increasing fibre will lower cholesterol, lower blood sugar, cut your risk of colon cancer, and is linked to extended longevity.

Most cereals, fruits and especially vegetables are fibre rich, and baked beans on wholemeal toast has just become a major health food!

Get fibre from: Pears, whole grains, berry fruits, beans, peas, lentils, avocados, apples, carrots, artichokes, nuts, seeds especially chia, bananas, oats.

Food synergies

Food synergy is where two or more foods combine to give a health boost greater than if the foods were eaten separately.

For example:

- **Broccoli** and **Tomato**. A study in the *Journal of Nutrition* reported that prostate tumours grew much less in rats fed tomatoes and broccoli combined, than in rats which ate diets containing broccoli alone or tomatoes alone. Cooking or slow roasting tomatoes in olive oil liberates far more of the cancer protective lycopene than when tomatoes are eaten raw. Add garlic for a tripled synergistic effect.

- **Green Tea** with **Lemon**. Green tea is already rich in polyphenols called catechins (pronounced ka-teh-kins) that are linked to lower cancer rates. A study from Purdue University found that adding lemon juice – or just vitamin C – produced a 400% increase in the bioavailability of the catechins.

- **Banana Milk** shake. Milk is a rich source of calcium. Calcium is better absorbed along with inulin, the prebiotic fibre found in bananas. Add a spoonful of wheatgerm because wheatgerm contains vitamin E and zinc, which help repair cells and strengthen the immune system.

- Stir fry **Spinach** with **Olive Oil**. Swiss/rainbow chard, spinach and kale are high in lutein. Lutein, however, is better absorbed with fatty acids. So, cook these leafy green vegetables in a stir-fry with olive or avocado oil.

- **Apple** and **Strawberry** compote. Apples and blackberries contain quercetin and strawberries contain fisetin. The combination is a powerful killer of senescent cells. Similarly, ellagic acid, found in raspberries, has been shown to boost the ability of quercetin to kill cancer cells. So, breakfast with mixed fruit compotes.

- **Oat** porridge/muesli/granola with **Berries**. Oats contain oat beta glucans, which can help lower LDL cholesterol and prevent the build-up of arterial plaque. A study at Tufts University found that when vitamin C was added to the oat beta glucans, it significantly boosted their cholesterol-reducing and artery-protective effects. Blueberries, raspberries, strawberries and blackcurrants are great sources of vitamin C and other antioxidants.

We will be posting recipes on acceleratedlearning.com/delay-ageing that are designed to create food synergy.

- 17 -

THE COMPREHENSIVE PLAN TO DELAY AGEING

Jonathan Swift, the 17th century Anglo-Irish satirist and poet, wrote:

"Every man desires to live long; but no man would be old."

I believe we have seen how these two objectives can be made compatible.

Ageing <u>can</u> be delayed

A 2019 review article in *Nature*, the world's most cited science journal, sums up the state of play in ageing research:

"We are now entering an exciting era for research on ageing. This era holds unprecedented promise for increasing human health-span: preventing, delaying or—in some cases—reversing many of the pathologies of ageing based on new scientific discoveries[2]".

You can do that by:

1. Increasing repair rates for DNA and mitochondria
2. Clearing away senescent cells by encouraging autophagy
3. Turning on health-protective genes
4. Improving your sensitivity to insulin and nutritional intake
5. Reducing glycation (the cross-linking of proteins due to high blood sugar)
6. Maintaining telomere length
7. Reducing the loss of stem cells

8. Reducing the level of errors in the proteins you create

9. Supporting your immune system

10. Maintaining the optimum ratio of good to bad bacteria in the gut.

11. Reducing inflammation

The plan we have been developing can help you achieve all these, combining to create a comprehensive wellness road map.

THE COMPREHENSIVE DELAY AGEING PLAN

Best of the best healthy delicious sustainable foods

✓ Cutting down on foods that increase damage.

✓ Increasing foods that improve repair – foods with a high antioxidant and anti-inflammatory content and which help express healing genes.

Activity and exercise

✓ Incorporate an active lifestyle programme that promotes positive epigenetic change.

De-stressing and mindfulness

✓ Add daily mind-calming deep breathing and a weekly mindfulness session.

Reduce calories

✓ Enhanced Hara Hachi Bu

Add a health food supplement

✓ A comprehensive one with a full range and optimum level of age-delaying nutrients.

'Best of the best' foods

The food plan presented next will not only help combat the 10 ways in which you age prematurely; it is also based on the 'best of the best' researched diets.

We have already seen that the **Mediterranean Diet** includes many age-delaying elements, but so does the typical **Japanese Diet**. So, it combines elements from these two.

Then it incorporates the elements of the **DASH Diet** – which was developed by the US National Institutes of Health. DASH stands for Dietary Approaches to Stop Hypertension (high blood pressure).

Finally, it incorporates the best elements of the **MIND Diet**, which was developed by Rush University with funding from the US National Institute on Aging. MIND stands for Mediterranean-DASH Intervention for Neurodegenerative Delay.

The MIND diet not only incorporates the foods that best protect against heart disease, stroke and cancer; it also includes foods and nutrients that help reduce the risk of Alzheimer's.

The MIND diet also lists 5 'unhealthy' foods, which you must restrict to one serving a week. The 'unhealthy' foods are fried or fast food, red meats, butter or margarines/spreads, cakes and pastries and sweets. MIND does recommend a glass of wine a day.

The objective of the MIND diet is to lower dementia risk and the Rush University team worked with a group of 923 seniors for 10 years. The results showed that the diet lowered the risk of Alzheimer's by as much as 53 percent in participants who meticulously adhered to the diet[248].

My dad used to say: *"Moderation is all things – including virtue!"*

So, our recommended food plan is a "moderate" one – and one that is not difficult to fit into any lifestyle.

Mainly plant foods – the reducitarian concept

The *Delay Ageing Food Plan* has a lot of plant foods but does not demand you become a vegetarian or vegan. More what has been termed a 'reducitarian' – eating mainly plant protein but with fish and some meat.

THE DELAY AGEING FOOD PLAN

GREEN foods	2 a day	Broccoli, chard, spinach, cabbage, kale, cauliflower, sprouts, rocket, asparagus, green peppers, peas, courgettes/zucchini, celery, leafy salads, mangetout /sugersnaps
RED foods	1-2 a day	Strawberries, raspberries, cherries, red grapes, red pepper, red onion, tomatoes, apples, chilli peppers
YELLOW and **ORANGE** foods	1-2 a day	Carrots, sweet potatoes, oranges, grapefruit, mango, apricot, pumpkin, banana, sweetcorn, squash
BLUE and **PURPLE** foods	1-2 a day	Blueberries, bilberries, blackberries, blackcurrants, black/concord grapes, raisins, plums, aubergine/eggplant, beetroot
WHOLE GRAINS	2 a day	Wholegrain bread, cereals, pasta, rice, oats, quinoa, buckwheat, rye, barley, etc
Fruit/vegetable **JUICE**	max 1 a day	Single or mixed fruits or vegetables (pure fruit/veg, no added sugar)
PREBIOTIC and **HIGH FIBRE** foods in addition to fibre in fruits, grains and vegetables	4 -5 times a week	Onions, garlic, leeks, artichokes, lentils, bananas, beans (kidney, haricot, borlotti, black-eye etc), chickpeas and products eg. dhal, hummus. Add occasional fermented foods, like sauerkraut, miso, kefir, tempeh.
MUSHROOMS	3 x a week	All eg. chestnut, portobello, oyster, shiitake, maitake, chanterelle, reishi

HERBS and **SPICES**	as frequently as you can	Turmeric, ginger, cayenne, chilli, curry powder, basil, thyme, black pepper, cinnamon, oregano, rosemary, nutmeg, sage, coriander, etc. They add health benefits and flavour, reducing need for salt.
SOYBEANS and products	1-2 times a week	Tofu, edamame, miso, textured soy protein, natto
NUTS and **SEEDS**	small handful daily	Nuts eg. walnuts, cashews, peanuts, almonds, brazil nuts. Seeds eg. flax (linseeds), chia, hemp, sesame, pine, pumpkin.
Oily **FISH** *(Optional)*	2-3 times a week	Salmon (esp. wild), herring, trout, anchovy, mackerel, sardines, pilchards. Occasional shellfish eg. prawns (for selenium, zinc).
MEAT *(Optional)*	1-2 times a week	Chicken, turkey, game, duck, lamb. You need about 1g a day of protein per gram of body weight. Organic/free range if you can.
EGGS *(Optional)*	up to 7 a week	Organic and free-range hens have better feed quality and lifestyle, and their eggs have higher nutrients.
DAIRY *(Optional)*	in moderation	Dairy contributes calcium, magnesium and vitamins, but these are also in vegetables. Choose organic milk and butter, real cheeses, especially green and blue, and plain 'live' yogurts.
Plant-derived **FATS** and **OILS**	as needed	Extra virgin olive, flaxseed or hempseed oils for salads and general cooking. Avocado oil for high-temperature frying.
DARK CHOCOLATE	2-3 squares daily	Cocoa flavonols are healthy and there's much less sugar and no dairy compared to milk chocolate.
DRINKS	frequent	6-8 glasses equivalent of water-based drinks ie. water, teas (green, black, chamomile, herbal), coffee. Moderate red wine (women 1 glass a day; men 1-2).

Commentary

The Food Plan is high in anti-inflammatory and antioxidant foods. There is a reason for the colour bands at the top. As we have seen, plants develop colours in their skin to defend themselves from oxidative damage by the sun and from insect attack. The colours signal their powerful antioxidant and anti-inflammatory properties. When you eat the plants, you inherit these benefits.

Each row represents the four colours you should have, at some meal, on your plate every day. It is like the advice to 'eat a rainbow'.

The colour coding idea comes from an excellent book called 'The Color Code'. The lead authors were Dr James Joseph, Director of the Neuroscience Laboratory at the Human Nutrition Research Center on Aging at Tufts University, with Anne Underwood, Executive Editor, Harvard Special Health Reports and Daniel Nadeau.

You can print a version out for the fridge door from acceleratedlearning.com/delay-ageing/food-plan.

Protein. You need about 1 gram of protein for each kilo of body weight as you get older. It's not very much. So, averagely,

- **MEN** (75 kg / 165 lb / 11 st 11 lb) need about **75g** (just under 3 oz) of protein a day.

- **WOMEN** (65 kg / 143 lb / 10 st 3 lb) need **65g** (about 2½ oz) of protein a day.

Protein should take up no more than a quarter of your plate. Excess protein activates the IGF-1 pathway and is linked to faster ageing. High protein diets, particularly those high in animal protein, are not healthy.

Dairy contributes protein, calcium, magnesium and vitamins, but these are also in vegetables.

Eggs. The liver adjusts the cholesterol balance so that eating eggs does not normally increase it.

Omega-3. Vegetarian sources include flax seeds, chia seeds, hemp seeds, walnuts. But this form of omega-3 is not as well absorbed as fish oil.

Yogurts. Choose plain 'live' yogurts, as flavoured yogurts have little fruit and high added sugar.

You can still use the Food Plan (with some exclusions) even if you are a vegetarian or (with more exclusions) if you are a vegan.

Good foods you are emphasising

Fruits and vegetables

Foods and nutrients work better in combination and home-made soups, stir fries and smoothies are an easy way to reach and exceed your 5-a-day fruits and vegetables.

Exploring combinations of fruits in fruit smoothies is fun, as long as you limit yourself to one a day, because they are high in sugars. No limit on vegetables, because vegetables (with the exception of carrots) have a very low glycaemic index, meaning they raise blood sugar only slowly, as they take time to digest.

Whole grains

The plan emphasises only whole grains. That's because a whole grain contains three main parts: the *germ*, or sprouting part of the grain; the *endosperm*, which contains the starch to feed the young sprout during its early stages; and the *bran*, which is the outer protective layer.

In a whole grain food, all three parts of the grain are retained; in a refined food product, like white bread, the germ and bran are removed. But that's exactly where most of the fibre, vitamins and minerals are. Whole grains also have a low glycaemic index and are more filling than refined grains.

Dark chocolate

You will probably be pleased to see that the plan encourages you to eat two or three squares of dark chocolate a day. The cocoa in dark chocolate contains phytochemicals that make it: *"a biologically active ingredient with potential benefits on biomarkers related to cardiovascular disease"* [249]. In other words, it's good for your heart.

What's more, a 2011 study in the *American Journal of Clinical Nutrition* showed that even your gut microbes like chocolate! The effect of feeding 70% cocoa chocolate was to increase bifidobacteria, lactobacilli and butyrate counts and, as a result, markers of inflammation were reduced. A study well worth reporting[250].

Sadly, for milk chocolate fans, the same results are not replicated. Milk chocolate contains between 26% and 35% cocoa solids, and the US brand leader Hershey contains just 11% cocoa solids. The rest of the cocoa is replaced in milk chocolate by a lot more (saturated) fat and sugar. A combination that is pretty addictive, as the fast food companies are all too well aware.

Unhealthy inflammatory foods you are largely excluding

You will be reducing foods that raise glucose levels sharply, are pro-inflammatory, or have an adverse epigenetic effect.

Refined sugar foods – sugary soft drinks, cakes, biscuits/cookies, confectionery/candy. These and refined starchy foods also raise blood sugar (glucose) levels fast, as they have a high glycaemic index, which is unhealthy and ageing.

Refined starchy foods – white-flour baked goods, white rice, crisps/chips, snacks – which also raise glucose levels.

Processed and fast foods with excess omega-6 oils like sunflower, corn and palm oil

Smoked or cured meats (eg. bacon, hot dogs) with nitrosamines linked to cancer.

The Delay Ageing Food Pyramid

You will have seen food pyramids before, but ours is inverted, with the foods to eat most of at the top, and foods to eat sparingly at the bottom.

Notice that the inverted foundation consists of **vegetables**, because veggies contain some of the most protective nutrients and fibre. But there is another reason. When University College London published their study confirming 9-10 portions of fruit and veg a day as optimal, they also showed that consuming vegetables had an even greater protective effect than eating fruit.

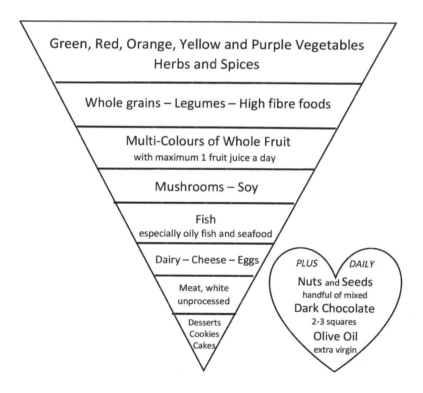

Why I recommend a daily nutritional supplement

A supplement is never a substitute for a healthy diet. But what a supplement can do is add <u>extra</u> levels of the nutrients that appear to have the biggest impact on health and longevity.

So, when I recommend supplementing certain nutrients, I am doing so as additions to an already very healthy diet. The main aim is not to correct deficiencies, but to reach an optimum level of nutrition – a big difference. Note they are natural – derived almost exclusively from plant sources.

As I also recommend a minor daily reduction in calories – Enhanced Hara Hachi Bu – a supplement is important. Most of the animal studies showing improved health and longevity from reduced calories included a supplement.

Healthy power nutrients in a daily supplement

In 2019, the Edith Cowan University in Australia issued an analysis of the diets of more than 53,000 Danish people over a 23-year period[251]. They found that people who regularly ate foods rich in flavonoids had a significantly lower risk of dying from heart disease or cancer.

That protective daily level was 500 mg of total **flavonoids** – so I believe that is the level you should aim for in a supplement. Based on the evidence, the following flavonoids are recommended:-

Curcumin – Green tea – Grapeseed – Lycopene

Lutein – Beta carotene – Bilberry/Blueberry

From evidence we have presented, other age-delaying nutrients are:-

Omega-3 – Soy isoflavones (especially Genistein)

Betaine – Glucosamine – Co-Enzyme Q10

These would provide you with a powerful combination of age-delaying nutrients to add to essential vitamins and minerals.

Combination is key

All these nutrients can be bought as supplements from retail or online sources and the combination is key. The aim is to create an anti-inflammatory supplement that helps counteract the original nine ageing pathways simultaneously, with a formula designed to have a synergistic effect.

The recommended combination is based not just on the studies we have been reviewing, but also on in vitro laboratory results.

These independent tests showed a significant anti-inflammatory effect – reducing six key inflammatory bio-markers to either zero or very low levels, specifically IL6, IL8, IL1-beta, prostaglandin-E2, TNF-alpha and isoprostane[252].

Inflammation is a key marker of ageing and disease, and the laboratory conclusion was that the combination could be classified as a 'nutraceutical', which the *International Journal of Preventative Medicine* describes as a *"natural product with a physiological benefit"* [253].

Full disclosure – my own mentor, Dr Paul Clayton

This book has mentioned a lot of researchers. But so far it has not mentioned my own mentor, Dr Paul Clayton.

I first met Dr Clayton back in 2000 when he was Chair of the Forum on Food and Health at the Royal Society of Medicine, specialising in the impact of nutrition on health.

Over time, he agreed to write a book, which became 'Health Defence' – a detailed explanation of what optimum nutrition is and how it can help prevent age-related illness. 'Health Defence', which I edited, was first published in 2002 and not only became a best-seller but was recommended reading on several medical courses.

In the book, Dr Clayton made a very strong case for supplementing even a healthy diet with the nutrients that I have just mentioned.

However, no such all-in-one supplement existed, and so with Paul Clayton as an advisor, my company developed a comprehensive supplement that matched the 'Health Defence' criteria. It includes 23 essential minerals and vitamins and a daily total of 500mg of flavonoids, polyphenols and carotenoids, plus genistein, CoQ10, betaine, brown seaweed extract and glucosamine.

It was this combination supplement that was tested. However, as already stated, you can get all the nutrients separately, and I have listed them all and the exact amounts on the acceleratedlearning.com/delay-ageing website.

Optimum is not just 'adequate' – vitamin and mineral intake

From the beginning, Dr Clayton focused on providing an 'optimum' level of vitamins and minerals as an important concept – a considerable advance on the 'RDA'.

RDA (Recommended Daily/Dietary Allowance) is a government-set value. Around the world, this is also referred to as DRI (Dietary Reference Intake) or NRI (Nutrient Reference Intake).

The RDAs of nutrients are established to prevent deficiencies and be *"sufficient to meet the nutrient requirements of nearly all (97%-98%) healthy people"*. Or where evidence is insufficient, *"to ensure nutritional adequacy."*

However, as you get older, gut and hormonal changes mean you may absorb rather less nutrition from food, and older people tend to eat less as their metabolism slows[254]. In addition, some medications decrease the body's ability to absorb nutrients and some older people are on multiple medications.

Consequently, the need for quality nutrition to counteract wear and tear with repair increases at the very time that nutrient intakes are often decreasing.

Add the fact that there has been at least some reduction in the nutrient content of some popular fruits and vegetables through intensive growing, and you have a situation where deficiencies can, and do, arise in people in their 50s and beyond.

These deficiencies are typically in vitamins B12, B3, D, C, E, K, magnesium and calcium. In these circumstances, 'adequate nutrition' does not mean 'optimum nutrition'. Therefore, I believe in taking a good vitamin and mineral supplement – as long as it is accompanied by a range of polyphenols and omega-3.

Additional nutrients for supplementation?

The food plan, with a comprehensive supplement, is designed as the foundation for delaying ageing and increasing your health-span. But we have mentioned a few other supplements along the way.

They include low dose (75mg) **aspirin**, a **probiotic** supplement and **1-3, 1-6 beta glucans** for the immune system. Should you add those?

Low-dose aspirin

The evidence is that low-dose aspirin does reduce the risk of heart attack and stroke by acting as a blood thinner, but it should only be taken with your doctor's advice, because it does carry a small risk of gastrointestinal ulcerations.

Probiotic multi

I now think there is an argument for older people to take a good-quality probiotic supplement (with a range of strains and a high CFU count) for a month, say, 2 - 4 times a year. The aim is to improve diversity in the gut, cut intestinal inflammation and increase the good to bad microbe ratio.

That's apart from using a probiotic for specific issues like IBS, candida and after an antibiotic – or possibly even to help alleviate mild depression.

1-3, 1-6 beta glucans

Finally, as we saw in Chapter 11, there is substantial evidence that 1-3, 1-6 beta glucans can support the immune system. I take it in the winter months.

Metformin?

Metformin, of course, is a drug, not a nutrient. But although we have mentioned it before as something that can work to delay ageing, which extends life in animal models and may lower cancer risk, it's not included in the final plan. That's because (a) our plan should reduce insulin levels and improve insulin sensitivity without the need for a drug; (b) most people will not be able to get hold of it as it's on prescription only; and (c) a 2017 editorial on Diabetes UK suggests that long-term use of metformin (at least by diabetics) may be associated with an increased risk of Alzheimer's. The American Diabetes Association has warned of the same possible link.

Should we classify ageing as a disease?

Some of the researchers on ageing are campaigning to have it classified as a disease, including David Sinclair, the Harvard geneticist. Old age, he says, is simply a pathology – and, like all pathologies, can be successfully treated.

Finance is the reason behind this stance – how disease is classified and viewed by public health groups helps set priorities for those organisations, like governments and healthcare providers, who control funds.

However, regulators have strict rules that guide what medical conditions a drug can be licensed to act on, and for what it can be prescribed and sold. Today, ageing is not on the list, although the World Health Organisation has added a new code for 'Age Related Diseases'. Sinclair says it should be on the list, because:

> *"If aging were classified as a treatable condition, then the money would flow into research, innovation, and drug development".*

It would, he says, be the *"biggest market of all"*.

Nevertheless, there are other academics who do not automatically take the drug route. Like Eline Slagboom, a leading researcher on ageing, who works at Leiden University Medical Center and The Max Planck Institute for Biology of Aging, and who is part of several EU ageing initiatives including the development of ageing biomarkers. She says:

"Viewing age as just a treatable disease shifts the emphasis away from healthy living."

Otherwise, the conclusion is:

"We can't do anything with anybody until they reach a point where they get sick or age rapidly, and then we give them medication."

I hope this book has made clear there are many effective ways to help counteract the hallmarks of ageing before we go down the route of pharmaceutical drugs. I am not necessarily against the idea of drugs that delay ageing, but I believe that food and naturally derived nutrients should be the first priority. They are not patentable and therefore not expensive.

Road paved with good intentions? Role of motivation

We are almost at the end of our journey towards what could be as many as 20 extra years of good health. But there is one element we can add. A bit of extra motivation.

Hopefully, your basic motivation will be the overwhelming evidence presented here that you can postpone ageing, which means that most people could extend healthy 'middle age' until well into their 80s. But it will require some effort, and effort requires, well, effort!

HEALTHY TO A HUNDRED
A Mind Map of the DELAY AGEING PLAN

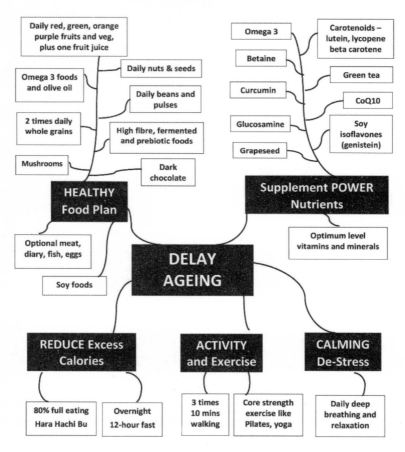

Tracking your Delay Ageing Plan

The tracking wheel below is developed from an original Buddhist concept. I have adapted it to include the eight lifestyle actions that are central to staying healthier, longer.

(You can download pdfs of this Tracking Wheel from acceleratedlearning.com/delay-ageing).

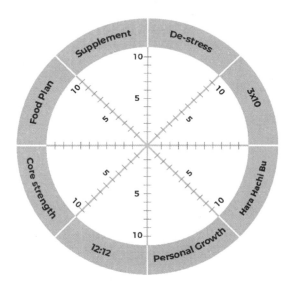

The eight lifestyle actions are:

1. **Food Plan.** Eat from all the categories over a week.

2. **Supplement.** If you are not reaching 9-10 fruit and veg a day, increase your polyphenol, carotenoid, vitamin, mineral and omega-3 levels with a supplement.

3. **De-stress.** Daily 5-minute mind calming, de-stress session.

4. **3 times 10.** Take a 10-minute brisk walk, 3 times a day.

5. **Hara Hachi Bu**. Follow the 80% eating path of Hara Hachi Bu to reduce calorie intake.

6. **Personal Growth**. What's the reason you get up in the morning? What are you doing that's new?

7. **12:12**. Twelve-hour overnight food break eg. 8pm to 8am.

8. **Core strength**. Do the four key exercises daily – plank, bridge, crunches, lunges.

Here's how you use the Tracking Wheel. You rate yourself currently on the scale within each segment. Then shade it in – like this:

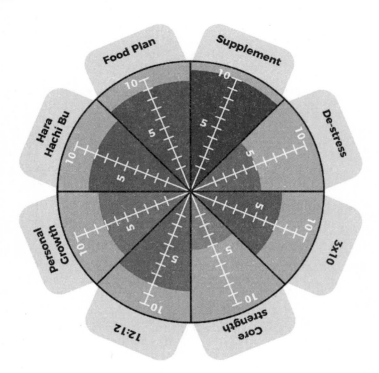

That's your start point. Does it look balanced? Are you low on some aspects? In a week's time, rate yourself again and see how the pattern has changed. Is it better balanced? Are your ratings going up?

Keep doing that every week over several months, because changing habits takes time.

How long? Phillippa Lally of University College, London has an answer. Her study suggests that it takes, on average, 66 days of doing something daily before a new more positive habit develops[255]. (The actual range was from 18 to 254 days.)

I recommend this Tracking Wheel because it makes getting and staying healthier and delaying ageing tangible, visible and actionable.

Updates and advice on healthy ageing

I have also set up a simple free service. Register at accceleratedlearning.com/delay-ageing and we will send you updates and advice on healthy ageing.

Sometimes it might include an update on work being done by one or more of the healthy ageing experts whose names have featured in this book. This is a young but fast-moving field. Other times it will include some tips and observations from people, who like you, are on the journey.

It will also contain recipes and it has mind-calming and mindfulness exercises and links to additional age-appropriate online physical exercises.

A final thought – use your extra healthy years to make a difference

Delaying ageing and achieving more years of extra, vigorous health is absolutely possible.

Camilla Cavendish calls it 'Extra Time' in her book of that same name. As an advisor to a former UK Prime Minister, she concentrates on the implications for society of a longer lived but healthier population.

You may have noticed that I have never used the term '*anti*-ageing' – until now.

First, because ageing chronologically is inevitable. It's biological ageing we are aiming to slow down.

Secondly, I don't view getting older negatively – only the adverse consequences of ageing. So, I like Camilla Cavendish's concept of 'extra time'.

Supposing that extra time was spent contributing to solutions to some of the pressing problems that face our communities and society? Using the experience that comes with age. There are certainly enough problems to choose from! But for each problem there is a definite role for a personal contribution, whether helping to tackle poverty, shrinking wildlife diversity, ecological restoration or global heating.

What one person does may not make a difference.

But what 5 million or 50 million do, makes a very, very big difference. So – what will you do with these 'extra years'? How could you make a difference?

It would be nice to think of an army of older people, who have left the pressures of paid work behind, focused on making the world a better place.

I'll close Part II with a favourite thought from Henry David Thoreau:

> **"None are so old as those who have outlived enthusiasm."**

PART III

HOT TOPICS

I have been writing on science, nutrition and health for many years.

So, I have included a section that expands on topics that I am most frequently asked about.

- 18-

Protecting yourself against Alzheimer's and dementia

None of the chapters so far have been about a specific disease. They are about delaying ageing, because if you delay ageing, you help prevent or delay age-related diseases generally.

However, there is one problem that concerns most people as they get older and therefore warrants more attention – the apparent inability of science to find a treatment for the form of dementia known as Alzheimer's Disease.

I say 'apparent' inability of science to find a treatment for Alzheimer's. Because, although the consensus amongst neuroscientists is that there is no cure for Alzheimer's, there is very real hope that the risk of this dreadful disease can be significantly reduced.

To understand how, let's start with the basics.

Alzheimer's – the facts

Alzheimer's results from a complex combination of genes, environment, and lifestyle. That's significant because you can influence your environment and lifestyle. And whilst you can't change your genes, you can influence how or if genes are expressed.

The single greatest risk factor for Alzheimer's is ageing. Most cases become overt after 65 years old. Therefore, if you delay ageing, you should delay or help prevent Alzheimer's.

The risk of developing Alzheimer's is 9% at age 65 and then increases steadily. At age 85 the risk is almost 33%.

Known risk factors, apart from ageing, include:

- Type 2 diabetes (which doubles the risk)

- Elevated glucose levels

- High blood pressure or hypertension

- High LDL cholesterol

- Obesity in middle age

- Lack of activity

- An unhealthy diet (high in saturated fats and sugar and low in vitamins, minerals and polyphenols)

- Smoking (increases the risk by 45%)

- High alcohol consumption

- … and former head injuries.

Apart from head injuries, you will see the strong connection between the risk factors for heart disease and for Alzheimer's. What's heart protective is head protective.

There is more than one type of Alzheimer's and typically at least two years elapse between the first symptoms and a diagnosis.

Brain scans show that the brain of an Alzheimer's patient can shrink by as much as 30%, as many neurons stop functioning, lose connections with other neurons, and die.

Average life expectancy after diagnosis is averagely 8 to 10 years.

Women have a higher risk of developing Alzheimer's than men, possibly due to changes in oestrogen levels and of course, on average, they live longer.

But women have the same risk of other forms of dementia. The most common of which is vascular dementia, where blood supply to the brain is restricted, often as a result of a stroke.

Older people of African and Central/South American heritage have almost double the risk.

Many people who develop Alzheimer's go through a prior stage called Mild Cognitive Impairment (MCI). Action taken at this stage gives the best chance of preventing the full disease[256].

Some researchers will add social isolation and depression as risk factors. But is depression association or causation?

Genetic susceptibility

There is a genetic component to Alzheimer's. The risk is greater if your mother had it, as opposed to your father.

Although there are over 20 genes that may increase the risk of Alzheimer's, patients with a gene called APOE4 are at particular risk.

At least 40% of confirmed Alzheimer's patients have at least one copy of the APOE4 gene variant. [Remember, you normally get two versions of any particular gene – one from your mother, one from your father.] Patients with two copies of the APOE4 variant have a 10-fold increase in risk.

It is no coincidence that APOE4 also appears to switch off or inactivate SIRT genes – which, as we have seen, are linked to longevity. And it activates a protein that promotes inflammation.

About 14% of the overall population in the UK and USA have the APOE4 gene variant. Massachusetts Institute of Technology (MIT) neuroscientists have found that APOE4 promotes the accumulation of the beta-amyloid proteins that cause the characteristic plaques seen in the brains of Alzheimer's patients[257].

Damage to neurons, synapses and neurotransmitters

One key fact needs an illustration.

A Neuron

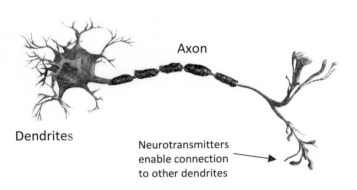

Your brain is made up of about 100 billion cells – called neurons. Neurons have branch-like ends called dendrites and dendrites from neighbouring neurons are linked together by synapses. Chemicals called neurotransmitters provide a 'bridge' between these synapses. This allows electrical signals – generated when we think – to communicate with other neurons in a vast neural network.

Neurons also have an axon – a tube through which the electrical signals that constitute thoughts flow. So, anything that damages a neuron internally, or damages or blocks a synapse, or damages a neurotransmitter will degrade the working of the brain.

In Alzheimer's, all three types of damage occur. To guard against Alzheimer's, you need to reduce neuron damage, ensure that any damage to synapses is balanced by the formation of new synapses and protect neurotransmitters.

Amyloid protein clumps

The defining characteristic of people with Alzheimer's is that they develop a sticky accumulation of proteins between brain cells. These protein clumps are made up of beta amyloid. The amyloid plaques collect between brain cells and disrupt cell function.

Amyloid proteins not only degrade the function of cells, they excrete toxins. These toxins damage neurotransmitters – and especially a key neurotransmitter called acetylcholine.

Tau tangles

Finally, instead of the proteins inside the brain's axon being normal, they develop tangles, called tau tangles. These tangles block and degrade electrical signals and thoughts, essentially, become confused.

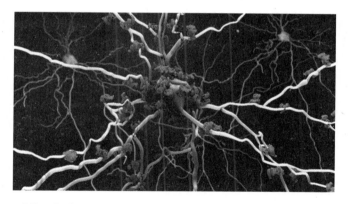

A unified theory of Alzheimer's?

All this means that Alzheimer's research has focused on either trying to clear amyloid plaque, preventing tangles, reducing neurotoxicity and repairing neurotransmitters. Yet, although billions have been invested in Alzheimer's drugs, nothing has worked.

Or has it?

Amyloid clumping is a response to damage in the brain

Earlier in this book we saw that acute and chronic inflammation are very different. Acute inflammation is the temporary result of your immune system zeroing in on short term damage – a cut, infection, a parasite or an insect bite. So, it's a response to damage.

Immune cells flood the infected area to defeat the pathogen and a by-product of that activity is inflammation, a reddening of the skin and often swelling. Once the immune cells have done their job and cleared the damage, the inflammation dies down.

But sometimes the damage is not completely cleared, which leaves a continuous level of internal inflammation 'grumbling' away. That is chronic or low-level continuous inflammation. We've already seen that this type of inflammation is a key driver of ageing and many age-related diseases.

The problem is that chronic inflammation is undetectable outside tests that only a doctor can make. They include a blood test for C-reactive protein (CRP), which is a direct measure of inflammation.

A second inflammation test looks for high homocysteine levels. Yet another test is for HbA1c levels. You will remember that HbA1c is a measurement of blood sugar, and it assesses damage to red blood cells. It tells us that high blood sugar levels are inflammatory.

Just as inflammation is a response to damage, leading neuroscientist Dr Dale Bredesen believes that amyloid, too, is a response – in this case to damage in the brain. But too much amyloid, like too much inflammation, can become a chronic problem.

So how do we prevent the brain damage that underlies Alzheimer's – the accumulation of amyloid protein, the tau tangles and loss of brain connections? Here are 15 ways to help prevent the shadow Alzheimer's falling on your family.

Preventing brain damage that causes Alzheimer's

1. Damp down inflammation

There is overwhelming evidence that inflammation in the brain contributes to Alzheimer's[258].

Keeping it as simple as possible, faced with damage in the brain from toxicity or amyloid, a part of the immune system for the central nervous system becomes overactive. It releases signalling molecules called cytokines which are pro-inflammatory. This inflammation then triggers the release of more cytokines in what becomes a vicious, damaging cycle.

Professor Edward Bullmore, in his 2018 book titled 'The Inflamed Mind', goes further. He states that inflammation in both the brain and the body is a root cause of depression, as well as Alzheimer's.

The answer is to ensure that your diet includes plenty of anti-inflammatory foods – the foods featured in the Food Plan.

Some of the best anti-inflammatory supplements include omega 3, curcumin, lutein, lycopene, beta carotene, green tea extract and grapeseed extract. Supplements we have already recommended.

In contrast, pro-inflammatory foods to avoid include fried foods, trans fats, refined carbohydrates where most of the fibre is removed (like cakes, white bread, biscuits/cookies, pasta), sugar, animal fats and processed meats.

One driver of inflammation is 'leaky gut syndrome'.

Leaky gut syndrome

The lining of your intestines needs to be permeable to allow nutrients, that have been extracted from food in your gut, to pass through the gut lining into your bloodstream. From there, they are dispersed and used throughout the body.

This dispersal is done at tiny junctions in the gut wall. Normally these junctions are tight enough to keep any particles of undigested foods out. But if the junctions are less tight than they should be, small fragments of undigested foods can pass into the bloodstream, where they are recognised by the immune system as 'foreign invaders'. This triggers an inflammatory response, which can contribute to Alzheimer's risk – and is implicated in auto-immune disease.

What causes the junctions to become damaged and 'leak'?

- Chemicals like pesticides

- Excess sugar, excess alcohol

- Chronic stress

- Several medications including NSAIDs (non-steroidal anti-inflammatory drugs like aspirin, ibuprofen, celecoxib), acid-reducing drugs and antibiotics

- Gluten can also trigger leaky gut in patients who have intolerance or sensitivity to gluten

2. Prevent insulin resistance

This is a largely a function of not over-eating and reducing sugar intake – because both increase glucose levels in the blood, so that your insulin response is over-activated.

Recent research adds an important new observation. Once insulin has done its job of lowering blood sugar, the body needs to rid itself of the excess insulin. It does this through an enzyme called insulin-degrading enzyme or IDE[259][260].

Now IDE also degrades amyloid, which is what you want to happen. But if IDE is having to work hard to degrade insulin, it struggles to degrade amyloid at the same time, as it is a competing function.

So, by keeping your blood sugar levels down, you free up IDE to tackle amyloid.

There is yet another reason to keep your blood sugar levels down – it reduces glycation. You remember that glycation happens when sugar bonds to proteins or fats and tissues become cross linked? Well, glycation is a factor in Alzheimer's[261].

Elevated blood glucose also damages blood vessels – in the body and brain. Blood vessel damage causes impairment to eye retinas in diabetics and causes progressive brain damage and shrinkage which leads to problems in learning, memory, motor speed, and other cognitive functions[262].

3. Prevent nutrient depletion

Unlike many cells in the body, which are fairly short-lived, neurons have evolved to live a long time – over 100 years in humans. Consequently, neurons must constantly maintain and repair themselves.

You need to ensure optimum, not just adequate, nutrition to support the maintenance and repair of your brain cells, synapses and neurotransmitters as you grow older. Especially as you absorb somewhat less nutrition from food as the years go by[263].

The adult brain is able to reorganise its vast network of neurons by forming new connections – which is known as neuroplasticity. Neuroplasticity means the brain continues to generate new neurons and synapses and connections throughout life, helping the brain heal from injuries. It's why stroke victims, who have suffered brain damage, can learn to walk and talk normally again.

An optimum level of vitamins and minerals and polyphenols is the priority to keep your brain functioning well.

Critical vitamins and minerals for brain health

Vitamin E, which protects brain cell membranes

B vitamins – especially folic acid (B9), B12 and thiamine (B1), which is essential for memory formation.

Vitamin C, which is a key antioxidant and helps protect brain cells from free radical damage.

Zinc and Magnesium – deficiency is associated with an increased risk of Alzheimer's.

Selenium is yet another mineral that you need in optimum amounts, because it not only helps mop up excess neuron-damaging free radicals, but it supports glutathione in promoting synapse health.

Betaine – an amino acid, not a vitamin or mineral. But it is an important nutrient in helping reduce the risk of excess homocysteine levels that can lead to Alzheimer's[264]. Homocysteine is linked to meat eating, and high levels are recognised to be a risk factor for heart disease as well as Alzheimer's.

Avoid mercury

What you definitely do not need is mercury. Workers who were making felt for hats in the 18th and 19th centuries were heavily exposed to mercury, which they used in the process. They frequently suffered memory loss, depression and insomnia as a result. This was referenced in 'Alice in Wonderland' by the Mad Hatter. Long-lived fish like tuna and shark accumulate mercury, so it is best to avoid them.

4. Boost brain-derived neurotrophic factor – BDNF

BDNF is one of the key proteins that helps support the survival and repair of existing neurons and the generation of new neurons. It also supports the healthy functioning of neurotransmitters.

High levels of BDNF are linked to improved learning and a reduced risk of Alzheimer's. In contrast, low levels of BDNF are associated with both Alzheimer's and Parkinson's diseases. Significantly, the protein also plays a role in the regulation of glucose[265].

You will not be surprised to learn that BDNF levels normally decline over time. But there are ways to keep your BDNF levels as high as possible.

Omega-3

A diet high in omega-3 is especially important, as a comprehensive review at the University of California Medical Center confirmed that omega-3 increases BDNF levels. Omega-3 has further important effects on brain health because it supports mitochondrial function, helps form brain cell membranes and increases glucose metabolism. Conversely,

"... a diet low in omega-3 results in decreased learning and memory".

Omega-3 also supports the formation of bone and cartilage and expresses genes that suppress tumour formation and inflammation.

A recent study in the journal *Nutrients* found that reducing sugar intake and supplementing with omega-3 and curcumin *"improves cognitive function in healthy ageing humans"* and working memory. The combination appears to work by simultaneously reducing inflammation and oxidative damage and supporting the formation of neurons[266].

Vitamin D

Vitamin D supports the development and maintenance of synapses. In winter, when there is so much less sun, a vitamin D supplement is recommended at a level of 2,000 IU a day.

Curcumin from turmeric

Curcumin not only also helps inhibit the formation of amyloid; it boosts BDNF levels, as do the antioxidant-rich polyphenols in berry fruits like blueberries, bilberries, grapes and grapeseed.

Vitamin E seems equally important in maintaining BDNF levels[267].

Reduce calorie intake, particularly from fat and sugar

Higher than average energy intake – AKA eating too much! – is associated with an enhanced risk for both Alzheimer's and Parkinson's diseases. Obesity in middle age is correlated with an extra risk of dementia – especially 'apple obesity' where fat accumulates around the midriff[268]. So reducing your calorie intake as recommended in Chapter 15 is important.

Animal studies show that a diet high in fat and sugar significantly lowers BDNF levels. And a cohort study showed that people who ate a low-calorie or low-fat diet had a significantly lower risk for acquiring neurodegenerative diseases than those who maintained a high calorie intake.

5. Reduce exposure to toxins and infections

If the production of amyloid is, at least in part, a response to toxins, then reducing exposure to toxins will mean the brain has less reason to create amyloid. Lessen your risk of neurotoxins by:

- Avoiding long-lived fish like tuna and shark, which are known to accumulate levels of mercury, which is a neurotoxin[269].

- Reduce your exposure to residual pesticides in foods. I am anxious not to over-exaggerate these risks, as some authors do. But I do try to ensure that I eat organic versions of what are called the *Dirty Dozen*[TM].

Since a main reason to buy organic is to avoid pesticide residue, you need to know which conventionally grown produce have the most pesticide residues on them – and which have the least. That's what the *Dirty Dozen*[TM] list tells you, and I have reproduced it at the end of the chapter.

Glyphosate-based herbicides (weedkillers), especially, have been shown to cause DNA damage and can act as endocrine disruptors in human cell lines. The US Department of Agriculture found that 98% of non-organic strawberries, spinach, peaches, cherries and apples tested positive for pesticide residue.

6. Counteract environmental pollution

Some exposure is inevitable, but a 2017 study[270] in the grandly named journal *Environmental Science and Research International* shows that certain plant-based food ingredients in green tea, onion, garlic, coriander, and turmeric/curcumin: *"offer detoxification, therapeutic and preventive effects against the poisoning effects due to these (environmental) pollutants."*

7. Protect against toxins produced by moulds

When the brain is threatened by mycotoxins – mould toxins – a response appears to be to produce amyloid.

Cruciferous vegetables contain phytochemicals called sulforaphane and glucosinolates that help counteract the toxic effect[270]. *Phyto* means plant.

Cruciferous vegetables include broccoli, cabbage, cauliflower, kale, brussels sprouts, watercress and daikon. Daikon is little known in the West, but is the most popular vegetable in Japan.

8. Reduce gastric reflux medications

Certain medications for acid reflux, heartburn and indigestion – called proton pump inhibitors – can increase the risk of stroke and dementia. A study on 74,000 people aged 75 and older found a 44% increase in dementia in people who took them, over those who did not[271].

Although the researchers pointed out it was an association and not necessarily a cause, we do know that gastric reflux medications can impair the absorption of key brain health nutrients like B12, zinc, and magnesium.

This negative interaction between drugs and nutrients is not uncommon when many older adults also take multiple drugs at one time, a situation known as polypharmacy.

In the USA, an estimated 44% of men and 57% of women older than age 65 take five or more medications. 12% of both men and women in this age group take 10 or more medications a day!

9. Support healthy hormone levels

The role of hormones in Alzheimer's is controversial, because so much of the research appears contradictory. Here is what we do know.

Hormones support the development and maintenance of healthy synapses and therefore brain function. Key hormones for brain function are oestrogen, progesterone and testosterone.

Oestrogen

Animal studies show that oestrogen confers a protective effect on neurons[272]. It also appears to increase the number of connections in the hippocampus, a part of the brain that is important for memory and learning[273].

Both men and women produce oestrogen, but as it is the principal female sex hormone, women have more of it. When women go through menopause, however, their bodies stop producing as much oestrogen.

Alzheimer's disease occurs more frequently in women than in men – 65% of sufferers are women. It is possible that the reduction of oestrogen levels after the menopause reduces their protection from brain damage. This may be linked to the fact that oestrogen can promote the body's own production of antioxidants and thus reduce free radical damage, including free radical damage to neurons. So younger women are protected.

This hormonal link to Alzheimer's has prompted many researchers to conclude that prolonging oestrogen supply though hormone replacement therapy (HRT) should help protect women against Alzheimer's[274].

This might be true. Most, but not all, studies indicate that women who initiated HRT therapy before the age of 55 have no higher risk of Alzheimer's than other younger women. On the other hand, women who started HRT after age 65 do have a higher risk – between a 9-17% increased risk[274].

So, what conclusions can we draw from all this? As regards HRT, the Alzheimer's Society states:

"Until there is better evidence, the potential benefits of HRT as a way to reduce the risk of Alzheimer's disease do not outweigh the potential risks of HRT, which includes an increased risk of certain types of cancer, heart disease and stroke".

Fortunately, there are ways to increase oestrogen levels naturally, because there are phytoestrogens – plant oestrogens – that naturally occur in plant foods.

These foods include flax seeds (the richest source), soy foods (which include isoflavones that are known to boost oestrogen levels), strawberries, peaches, apricots, prunes, dates, whole grains, peanuts and walnuts – and red wine.

Phytoestrogens have other benefits; they may also help reduce the risk of osteopenia (bone loss) and osteoporosis[275 276].

In addition, researchers have noted the significantly reduced levels of breast cancer in Japan where soy intake is high, but this may be an association with generally better diets, and with green tea consumption, rather than a direct cause.

The overall conclusion is that, in moderation, the consumption of phytoestrogens can reduce the risk of dementia and is safe, in food or supplement form. A large meta-study of phytoestrogens published by the Society for Endocrinology concludes:

> *"Current evidence would suggest that a high dietary intake of phytoestrogens does not increase the risk of breast cancer."*

Testosterone

Men appear to need to be less concerned about the link between hormones and Alzheimer's. Testosterone is converted into oestrogen inside brain cells and men continue to produce testosterone, the male sex hormone, throughout their lives. (As do women, but at much lower levels.) It may be the higher testosterone level that continues to account for men's lower Alzheimer's risk.

10. Increase autophagy (getting rid of dead cells)

We saw in Chapter 1 that autophagy helps clear senescent cells – and that includes senescent brain cells. It can also help clear amyloid before it begins to clump together.

11. Find time each day to de-stress

Stress raises cortisol levels that damage neurons in the hippocampus. People who are under stress produce less BDNF, which may be why people often say that when they are stressed, they cannot think clearly.

12. Ensure good oral hygiene

A 2019 study clearly shows that a bacterium called P. gingivalis, which causes gum disease, is frequently present in the brains of Alzheimer's patients[277] [278]. It may be a toxin that triggers the production of amyloid.

Brushing your teeth in the morning and evening is, therefore, an important part of brain health.

13. Increase your activity level

Brazilian and other researchers have shown that even short bursts of exercise can dramatically raise BDNF levels in the hippocampus – an area vital for learning and memory formation, thereby *"counteracting the mental decline associated with ageing"* [279].

We also know that sitting for hours on end without getting up and moving has a very bad effect on health generally. Some articles have classified continuous sitting, without taking a regular short break, as being as 'bad as smoking'. Whilst cigarette companies can justifiably, in my opinion, be termed mass murderers, to equate sitting with smoking is just a newspaper-created myth.

That said, sitting is bad for you. The dangers of continuous sitting include obesity, increased blood pressure, high blood sugar, excess body fat around the waist and abnormal cholesterol levels. Not surprisingly, prolonged periods of sitting also seem to increase the risk of death from cardiovascular disease and cancer.

Physical activity is also central to a healthy brain. A 2013 study showed that activity in older people improved cognition in precisely the area often most adversely affected in Alzheimer's – the hippocampus [280]. The hippocampus is critical for memory and the prefrontal areas are key to decision making. Dancing has also been shown to boost BDNF levels [280].

14. Improve your immune system's ability to destroy pathogens

Pathogens can be external bacterial or viral threats, or internal threats caused by a microbiome with a poor ratio of good to bad bacteria – dysbiosis. Either way, viral or bacterial damage in the brain can trigger amyloid.

The Delay Ageing Food Plan includes foods that support the immune system. But there is a good case to give it extra support in times of specific threat.

One of the few natural products that does enhance the immune system quickly is a supplement we have already discussed, called 1-3, 1-6 beta glucans.

15. Boost number of brain connections to replace losses

One of the puzzles of Alzheimer's is that autopsies show that some people had significant amounts of amyloid plaques and tau tangles – and yet had never displayed signs of dementia in their lifetime.

Somehow, they have made enough new connections and neural pathways to compensate for the ones lost or damaged. We know this is possible because it is the 'plasticity' or rewiring of connections that recovering stroke patients achieve.

A famous example of plasticity comes from a study of London taxi drivers by neurologists at University College London. They found that the average cabbie's hippocampus, a part of the brain primarily associated with memory and spatial navigation, grew significantly larger as a result of having to memorise the huge and complicated road map of London. Of course, satnav technology has probably put a stop to this phenomenon.

You can support your own neuroplasticity, or rewiring to create new connections, through sleep which resets connections, and through **continued learning of new subjects** and skills. Making music[281], creating art[282], and learning a language have all been shown to boost the formation of new connections. Learning new things is especially important for people who have retired early.

Neurogenesis

Contrary to previous belief, we now know that it is possible to create new brain cells, a process called neurogenesis. Researchers at Stanford University have found that certain foods support the formation of new brain cells. These include the flavonoids in berry fruits, cocoa (and dark chocolate), nuts, curcumin and leafy green vegetables.

Vitamin E, omega-3 and epigallocatechin-3-gallate (EGCG), the key phytochemical in green tea, all promote neuron formation in the hippocampus. The Stanford researchers conclude that:

"A vitamin-rich, low-fat diet aids neurogenesis in experiments with rodents, and a low-calorie diet mitigates the effects of neurogenerative disease in mice."

Get enough sleep

It is not often that a whole new biological system is discovered in the body. But that is what happened in 2012 when brain researchers discovered the **glymphatic system**[283].

The glymphatic system – so named because it involves glial brain cells and has similarities to the lymphatic system – is like a layer of 'piping' that surrounds the brain's existing blood vessels. Its role is to drain waste products and toxins from the brain.

Jeffery Iliff was on the team that discovered the glymphatic system and noted:

"If the glymphatic system fails to cleanse the brain as it is meant to, either as a consequence of normal aging, or in response to brain injury, waste may begin to accumulate in the brain. This may be what is happening with amyloid deposits in Alzheimer's disease."

The research team confirmed that in mouse experiments, more than half the amyloid removed from an Alzheimer's brain is removed via the glymphatic system.

So, ensuring that your glymphatic system is working well is a key brain protector – and it does most of its work at night while we are asleep. This is so significant that the researchers have even suggested that this may be the key reason that sleep is restorative.

Conclusion? For brain health, ensure you get that 7-8 hours of sleep a night.

Can Alzheimer's be reversed?

So far, we have identified no less than 15 ways that the risk of Alzheimer's can be substantially reduced. But at least one neuroscientist believes it can be reversed, if caught early enough.

Enter Dr Dale Bredesen. He is an expert in neurodegenerative diseases and has held senior positions in neurology at the University of California. He founded the now famous Buck Institute for Research on Aging in 1998.

Over the last few years Dr Bredesen and his team have developed a protocol – a programme for Alzheimer's patients, called the 'Recode Protocol'. They not only claim it can halt the disease, they also claim that symptoms can be reversed, if the patient is treated early enough. Dr Bredesen details this programme in a book called 'The End of Alzheimer's'.

In the book, he gives examples of people who have halted their Alzheimer's and there was a small published study in 2014 on 10 patients that details these improvements[284]. A further trial commenced in June 2019 and is due to complete in 2020.

The book is well worth reading if you suspect a loved one may have the early signs of dementia. However, I suggest that he has been badly served by his publisher. The title may stop you in the bookshop aisle, but the 'The End of Alzheimer's' seems to be overclaiming.

Which means that, although Dr Bredesen has supporters amongst his peer neuroscientists, he also has his critics.

However, that does not mean that Dale Bredesen's concepts are wrong. Just that they will not work for everyone. Nor should we expect them to: Alzheimer's is a highly complex disease, with many contributory factors. These will affect individuals, with their unique genomes and biomes, in different ways.

It is precisely because of this complexity that Dr Bredesen believes trying to find a single 'magic bullet' drug that will tackle the multiple causes of Alzheimer's is very unlikely to succeed.

You will see parallels for ageing. Since there are so many contributing factors for ageing, it is very unlikely that any one drug will ever work to halt or reverse ageing. Which is precisely why we are exploring a multi-factorial approach.

Alzheimer's 'recoding' is complicated

Dr Bredesen's Recode Protocol asks that you consider no less than 36 factors that could be contributing to Alzheimer's. It encourages a patient to have tests for key markers followed by a year's clinical supervision on a personalised plan at an approved clinic.

The main criticism from fellow researchers is that this is too complicated for many people to follow unaided, which may well be true. The critics also complain that there are so many parts to the protocol, it is impossible to tell which part is working.

But does it necessarily matter? Alzheimer's is a very complex, multi-factorial disease and only a multi-strand approach is likely to be effective. In fact, a reductionist wish to create a single magic bullet drug is the reason why Alzheimer's pharmaceuticals have all failed so far.

Meantime, the 15 strategies above will not reverse Alzheimer's, but, coupled with the full *Delay Ageing Plan*, it will give you the best possible chance of keeping it at bay.

The 'Grain Brain' – myth?

Earlier, we saw that inflammation is a driver of Alzheimer's, and in some people, gluten triggers inflammation. So, we should not leave this chapter without mentioning the American Dr Perlmutter and his best-selling book 'Grain Brain'.

In the book, Dr Perlmutter, described in Wikipedia as a 'celebrity doctor', makes some uncontroversial points. He points out the link between Alzheimer's and diabetes, high blood sugar levels, inflammation, and inactivity – all of which are well documented.

He also states that eating grains that contain gluten can trigger inflammation in body and brain in people with gluten intolerance and with gluten sensitivity. Which is also true.

So, gluten may well be a potential contributory factor in the development of what he calls 'brain fog' and depression, and even Alzheimer's, for gluten sensitive people.

Since there seem to be about 7% of people who have gluten intolerance or sensitivity, in a 330-million US population, that would be about 23 million people. The equivalent number in the UK would be about 4.6 million. So, it would not be hard to find many Americans or Britons who have followed his advice to cut out gluten and benefited, and Dr Perlmutter's book and website have plenty of such cases.

It is when he extrapolates from the fact that inflammation is a factor in Alzheimer's, and gluten can trigger inflammation, that therefore <u>everyone</u> should avoid gluten – and almost all carbohydrates – that Dr Perlmutter has attracted mainstream medical criticism[285].

Especially when he then recommends replacing those grains and carbs with fats, to the point where some 60% of your intake would be fats.

I suggest we stay with the hundreds of studies that support a Mediterranean or Blue Zone style diet to prevent Alzheimer's.

A brain booster supplement? Bacopa monnieri

Are there any supplements that could help in specifically supporting cognitive function?

I think one may be worth trying – if you have a specific reason to be concerned about declining cognitive powers. It has been subjected to the 'gold standard' in scientific research, which is a randomised, double blind, placebo-controlled trial.

The supplement is a traditional herbal medicine called Bacopa monnieri, also known as water hyssop, brahmi or herb of grace. Bacopa monnieri has been used for centuries in Ayurvedic medicine as a memory and learning enhancer and stress reducer.

Bacopa monnieri contains active compounds called bacosides, which have been shown to regulate the level of important neurotransmitters. Its main mechanism is to help promote neuron communication, and it does this by apparently increasing the growth of neuron nerve endings called dendrites.

A key trial was conducted at the National College of Natural Medicine in Portland, Oregon. In the Portland test, 54 people with a mean age of 73 were randomly assigned to either bacopa (300mg) or a placebo over a 12-week period. The bacopa group had significant improvements in memory scores and concentration and interestingly a decreased heart rate[286].

The findings of this study are supported by at least five other trials in humans totalling over 600 people.[287 288 289 290]

Each of these trials concluded that bacopa can improve attention, cognitive processing, working memory and retention in older people and may improve higher-order cognitive processes. A 2011 animal trial may indicate why – it showed that bacopa increases dendrite connections[291].

Bacopa monnieri has other positive properties. In animal and test tube studies, it shows anti-inflammatory effects comparable to two non-steroidal anti-inflammatory drugs commonly used to treat inflammation.

The reduction in anxiety and reduced heart rate effect also shown in some of the referenced trials is probably because bacopa is classified as an adaptogen – meaning that it increases your body's resistance to stress. Whilst improvements in cognition are consistent across all the trials, the reduction in anxiety, however, was less marked and less consistent.

If you are experiencing more than a few trivial episodes of forgetfulness – which is normal with getting older – bacopa is worth trying.

According to the studies, you should notice an improvement in mental sharpness after a 12-week trial, but if not, simply discontinue. It is not an expensive supplement and it is generally safe, although pregnant women should avoid the herb and anyone on medication should speak to their healthcare provider.

Should you train your brain?

There are many 'brain training' courses and apps available, some free. But do they work? The jury seems to be out[292].

An article in *Medical News Today* by Walter Boot, Professor of Cognitive Psychology at Florida State University, makes the very good point:

> *"Few people care much about improving their score on an abstract computerized brain training exercise. What is important is improving their ability to perform everyday tasks that relate to their safety, well-being, independence and success in life.*

> *"Over a century of research suggests that learning and training gains tend to be extremely specific. Transferring gains from one task to another can be a challenge ... playing brain games may only make you better at playing brain games."*

Bobby Stojanoski, a research scientist in the Brain and Mind Institute at Western University, sums it up well:

"If you're looking to improve your cognitive self, instead of playing a video game or playing a brain-training test for an hour, go for a walk, go for a run, socialize with a friend. These are much better things for you."

Chapter SUMMARY

- Follow the ***Delay Ageing Food Plan*** which features foods that have a powerful anti-inflammatory and antioxidant effect.

- Ensure any supplement is more than a simple vitamin and mineral pill and includes omega-3, curcumin, lutein, zeaxanthin, lycopene, green tea and grapeseed extracts.

- Reduce sugar rich and fat rich foods.

- Destress daily and follow good oral hygiene.

- Engage with an activity plan and keep learning!

- I think there is enough evidence that bacopa monnieri can improve attention and memory for people with Mild Cognitive Impairment to consider trying it for 90 days – but only with your doctor's permission. And note that bacopa may increase the effects of the anti-depressant amitriptyline.

- Reduce pesticide risks by buying organic, or growing your own, of the *Dirty Dozen*™ fruits and vegetables. Full details are overleaf.

FRUITS AND VEGETABLES
EAT ORGANIC WHERE YOU CAN

You will see that many of the Dirty Dozen™ fruits and vegetables have either no skin, or a thin skin, which is normally eaten.

That's why they are often best bought organic, since they have a higher risk of cross-contamination and pesticides are more easily absorbed.

It is generally safer to eat non-organic fruits and vegetables with a thick peel, such as avocados or bananas.

By buying organic versions of the 'dirtiest' produce ie. those with the most contamination, and by buying conventionally grown versions of the cleanest produce, you get the best value for money. (Assuming you can't grow your own or buy all organic.)

In terms of environmental impact, organic growing methods are much safer for all plants and animals, protecting wildlife and nature as well as just our own human health.

THE DIRTY DOZEN™

EWG publishes a shoppers' guide every year – the figures quoted are from March 2020. DEFRA, the UK's Department for Environment, Food and Rural Affairs, also did a survey. The % in brackets is the percentage of samples that DEFRA found contained residues in the UK, although they did not cover all the items.

		USA samples with pesticides per March 2020 list	UK data
1	Strawberries	**99%** had at least **1 pesticide**. **30%** had over **10** separate pesticides.	
2	Spinach	**97%** had pesticide residue, including neurotoxic permethrin	
3	Kale	**60%** had residue of pesticide DCPA, a possible carcinogen	DCPA was banned in the EU in 2009.
4	Nectarines	**94%** had at least **2** pesticides	
5	Apples	**47 different pesticides**, including 6 known or suspected carcinogens	UK **80%** with residues
6	Grapes	At least **5 pesticides** on average	UK **87%** with residues
7	Peaches	At least **4 pesticides** on average	
8	Cherries	At least **5 pesticides** on average	UK **95%** with residues
9	Pears	**5 or more pesticides** in more than half	UK **94%** with residues
10	Tomatoes	At least **4 pesticides** on average	
11	Celery	**95%** with pesticide residue	
12	Potatoes	Chlorpropham, **pesticide** used to control sprouting during storage, predominates on US potatoes	Chlorpropham was banned in EU in 2019.

- 19 -

Why does food and nutrition advice keep changing?

Americans consume an average of 26 kg (57 lb) of sugar a year, the average weight of a fully-grown female labrador dog! UK consumers (21 kg /47 lb) are not far behind. A bad interpretation of nutritional research is partly to blame – as we shall see.

Here are some of the confusing claims that have been made.

Coffee is bad for you.	-OR-	**Coffee** is good for you.
Eggs are unhealthy.	-OR-	**Eggs** are a superfood.
Cholesterol is a cardio-danger.	-OR-	**Cholesterol** has little impact on heart health.
Margarines/spreads are bad for you.	-OR-	**Butter** is worse.
Dairy is good for you.	-OR-	**Dairy** is unhealthy.
Vitamin supplements are a waste of money.	-OR-	Take a **multivitamin**.
Meat is bad for you.	-OR-	**Meat** is a great source of protein and vitamin B12.
A glass of **red wine** has health benefits.	-OR-	There's no safe level of **alcohol**.

Avoid fat ... carbohydrates ... grains ... gluten ... dairy.

Nutritional values of food have declined over the last 50 years.

Who's right? Who's wrong? And why *is* dietary advice so often contradictory? There are six reasons.

Why dietary advice is so often contradictory

1. Research findings can be, and often are, selectively used by the media in click-bait headlines and by publishers of diet books to hype whatever fad they are pushing. Headlines need to be either exciting, warning of danger, promising instant results, or controversial, to grab attention.

2. Perhaps most importantly, because we do not all respond to food (or nutrients or drugs) the same way. In contrast, laboratory rats and mice, which are often used in food trials, are selected to be identical. In humans, variations in genomes, in microbiome make-up, and in lifestyles, can produce different results from the same diet, nutrient or food.

3. Science moves on, and new research rightly outdates old. That's progress.

4. You need to check who sponsored or financed the study. Coca Cola have an Institute for Health and Wellness staffed by dieticians and doctors. Would you expect a study financed by that body to find that fizzy soda drinks had a negative effect on health?

   Moreover, a study done with 20 people is not going to be as reliable as one done with 200. Equally, a study cited by 10 other researchers is not likely to be as robust as one cited by 400 academics.

5. Research studies have typically involved more men than women. This not only ignores the fact that the reaction of women to drugs and nutrients can be different, but it can have and has had a damaging effect on women's health.

   For example, research on heart disease has consistently under-represented women and several conclusions, largely based on research on men, do not apply equally to women. Historically, this has adversely affected the effectiveness of cardiovascular treatment for women[293].

6. There are several types of clinical study, and the findings for each have varying degrees of reliability, as we'll see very shortly.

You'll have noticed that I have tried to support any main statement in this book with a numbered reference. You can therefore check the details, level of significance, and which type of study is involved.

Despite this cautionary introduction, there <u>is</u> dietary and lifestyle advice that almost all experts can agree on.

The main types of clinical trial

Test tube research

The starting point for both drug and nutrient trials is test-tube or petri-dish research ('in vitro' = 'in glass' studies) on cells or tissue in a controlled laboratory setting. That enables the researcher to isolate the effect of a single molecule. For example, the anti-inflammatory effect of curcumin on a cell or the effect of vitamin D on cancer cells[294].

But what works on a cell may not – or may – work the same in a living being, comprising 37 trillion cells. So, the next stage is often …

Animal study

Fortunately, research is increasingly done with computer modelling, but animal tests are critical to some studies. Common test subjects include the *C. elegans* nematode worm, the *Drosophila* fruit fly and mice or rats.

You may be surprised that *C. elegans* nematodes have neurons, skin, muscles, gut and other tissues that are very similar in form and function to, and share many genes with, humans. Both have about 20,000 genes. Consequently, many basic hypotheses are tested on them first. In view of some human behaviour, you might not be so surprised that we share most genes with rats!

Animal experiments mean that researchers can again isolate the effect of a single food or nutrient on an animal subject's health or behaviour. Of course, what works for small creatures may not work when scaled up in humans. Nevertheless, many results of the effect of individual nutrients on organs like the heart or liver would not be possible without animal studies.

Randomised controlled trial

This is a clinical trial that randomly assigns participants to two or more groups. One gets the treatment, be it a drug or a nutrient, the other group gets a placebo; then the outcomes are compared.

The gold standard is a randomised, double-blind, placebo-controlled study where neither the researchers nor the patients know which group the molecule is being tested on. This stops unconscious bias[295].

An example would be a study comparing the administering of either zinc or vitamin A to treat children with acute respiratory infection[296]. [The zinc had a positive effect, the vitamin A did not.]

Meta-analysis

This is a study of studies – a statistical process that combines data from many different research studies on the same subject.

An example would be a meta-analysis reviewing the role of optimum vitamin K levels in treating osteoporosis, in helping prevent hardening of the arteries, in treating arthritis and, along with other nutrients, in helping defend against cancer[297].

Cohort study

This is a clinical research study in which people who currently have an illness or condition receive a treatment. They are followed over time and compared with another group of people who do not have the treatment. A problem is that the result could be due to factors other than the treatment.

Observational study

These are essentially surveys, and possibly the most famous example is the Framingham (Massachusetts) Heart Study which followed 5,209 adult subjects for over 70 years – from 1948 to the present day.

The objective was to see who developed heart disease and how their lifestyle was different from those who did not. By the 1960s, it had already shown that cigarette smoking was a big factor in cardiovascular disease. It has subsequently indicated that high LDL cholesterol, hypertension, physical inactivity, obesity and uncontrolled stress are also factors.

Epidemiological research

This is similar to observational study. Epidemiologists study diseases in whole populations to determine specifically how, when and where they occur. The aim is to determine what factors are associated with diseases (the risk factors), and what factors may protect people against disease (the protective factors).

An example might be how the lifestyle of Blue Zone dwellers contributes to their longevity. But when many observational studies rely on people recording what they ate and how often they exercised, there is room for error!

You can begin to see why contradictions arise. A test tube study may indicate a successful outcome, which is not borne out in animal trials. Or an animal trial throws up a positive result, whilst observational epidemiological research may indicate that the result on mice does not hold for humans.

For example, in Japan, there is high green tea consumption and lower incidence of breast and prostate cancer than in the West. But is that correlation (there is a relationship between the variables) or causation (the one causes the other)? Green tea is probably a factor, but soy consumption and a high rate of vegetable intake are also likely to be contributory.

Nutritional research interpretation

So how do we best interpret nutritional research? We should look for consistency and the magnitude of the effect. Do almost all the studies say the same thing and is the result statistically significant?

Are test tube results at the cell level confirmed by animal tests, which are then confirmed by human trials, which are then confirmed by observational studies? If so, the conclusion is likely to be robust.

It is this consistency that enables us to say that having substantial levels of polyphenols and flavonoids in your diet is a cornerstone of healthy longevity – as are physical activity, de-stressing and foods containing fermentable fibre. In contrast, high-sugar, high-fat processed foods are inflammatory and unquestionably bad for health.

Individual nutrients that pass the consistency test as positive for health include curcumin[298], green tea, grapeseed extract, betaine for heart health[299] and lutein for eye health.

The evidence for omega-3 is generally stronger for brain health than heart health, although the Harvard School of Public Health has stated[300]:

> *"There is strong evidence that eating fish or taking fish oil is good for the heart and blood vessels. An analysis of 20 studies involving hundreds of thousands of participants indicates that eating approximately one to two 3-ounce servings of fatty fish a week—salmon, herring, mackerel, anchovies, or sardines— reduces the risk of dying from heart disease by 36 percent".*

Combinations, not single nutrients, are key

The pharmaceutical industry thrives on 'magic bullets' – single drugs that target a specific point in the development of a disease. For instance, arsphenamine, trade-named *Salvarsan*, cured syphilis.

248

You run a clinical trial, show a direct cause and effect, and shareholders are happy.

But nutrition doesn't work like that. It is almost always the combination, the synergy, that creates the beneficial outcome. As long ago as 2004, the HALE project[301] showed that the Mediterranean Diet, combined with physical activity, moderate alcohol, and no smoking created an almost 70 percent reduction in chronic disease.

A later study in *Circulation* found it was the synergy between consumption of fish, fruit and vegetables in the Med diet that specifically produced a reduced risk of blood clots.

Cholesterol research caused confusion

For years, we were led to believe that cholesterol was a super-baddie. Here are the facts.

Cholesterol is part of almost all the cell walls in your body – you need it. It is an essential component in many vitamins and hormones. About 80% of your cholesterol you make yourself, as opposed to about 20% that comes from the diet. When you eat foods with cholesterol, your body normally makes less internally to compensate.

Lipoproteins – fats (lipids) joined to proteins in the blood – transport cholesterol around the body, where it is used in almost every cell to build cell walls. Some are high density lipoproteins (HDL), some are low density lipoproteins (LDL).

It's the LDL cholesterol that is harmful, particularly oxidised LDL cholesterol. It allows lipids to become attached to the walls of blood vessels, which leads to a build-up of plaque, which becomes hardened, narrowing or blocking arteries and potentially resulting in a heart attack or stroke.

In contrast, if your liver produces a lot of HDL, then lipids are transported around safely. This means that total cholesterol is not a good marker of heart health – but the ratio of HDL to LDL is.

An even better marker is called ApoB, which indicates the number of cholesterol-laden particles circulating in the blood – a truer indicator of the threat to our arteries than absolute cholesterol levels. But not every doctor will order a test for it. According to cardiologist Allan Sniderman of McGill University, they should, because not knowing the ApoB result *"could endanger your life"*.

All fats were lumped together as bad

Here's the skinny on fats.

Mono-unsaturated fats come mainly from olive oil, peanut oil, avocados, some nuts, and rapeseed (canola) oil. Of all the fats, olive oil has the clearest health benefits.

Omega-3 is a polyunsaturated fat found in fish oil and flaxseeds. It is anti-inflammatory and is a very healthy fat. According to a *Harvard Health* update, omega-3 fatty acids may help prevent and even treat heart disease and stroke. In addition to reducing blood pressure, raising HDL, they help lower triglycerides (blood fat) levels[302].

A large-scale review in 2015 shows that omega-3 is even important for healthy brain function[303].

Polyunsaturated fats are essential to the diet, as your body needs them, but cannot make them. Omega-6 polyunsaturated fat foods include nuts and seeds, safflower and sunflower oil and eggs.

It used to be thought that omega-6 fats promoted inflammation, but a report from the *American Heart Association*, written by nine independent researchers, confirmed that data from dozens of studies support the cardiovascular benefits of eating omega-6 fats as well as omega-3[304].

Nevertheless, most people consume far more omega-6 fats than omega-3 and it is advisable to try to reduce omega-6 and increase omega-3 in your diet to achieve a better balance.

Trans fats or **hydrogenated fats** are found in processed or fried foods and are unequivocally bad for you.

Saturated fats come from meat and dairy products and have been the villains of the piece. Until recently.

A meta-analysis of 21 studies, published in 2019 and covering 347,000 people, concluded that there was not enough evidence to say that saturated fat increases the risk of heart disease. It has caused a lot of controversy. As eating fat does not increase insulin levels, and since high insulin levels increase cardiovascular risk (and other risks), adding some saturated fat to the diet may not be unhealthy. However, the safer conclusion is to mostly replace saturated with polyunsaturated fat, like omega-3 oils, which should reduce risk of heart disease, as polyunsaturates are cardioprotective.

Olive oil is the healthiest fat

So where does all this rather confusing evidence leave us? With olive oil as a clear winner.

Olive oil contains an astonishing array of polyphenols. It really is worth paying for high quality extra virgin olive oil, which is cold pressed and contains more polyphenols. These polyphenols are at least partly responsible for the results of a very large-scale (7,500 person) study called the PREDIMED study. The PREDIMED study assessed two groups both at high risk of heart disease.

One group was put on a low fat diet, advised to avoid meat, olive oil, nuts and dairy, unless this was low fat dairy. The second group was put on a Mediterranean diet, characterised by a high intake of whole grains, legumes, vegetables, nuts and fruits, fish, poultry and dairy (usually cheese and yogurt), wine and olive oil. A lot of their cooking included garlic, onions and tomatoes with olive oil. This diet is high in fibre, too.

A sub-group of the Mediterranean diet group was told to consume and cook with an additional bottle of olive oil a week, and a second sub-group was told to add more nuts each week.

The trial continued for over four years until it was stopped – because the results were so clear that it would have been unethical to continue. The Mediterranean group had 30% fewer incidences of heart attacks, strokes, and breast cancer. Their cholesterol levels and blood pressure were also lower[305]. They even reduced their average waist measurements, which indicated that they had lost some visceral (abdominal)fat.

Subsequent research on olive oil (posted on the NHS website) indicates that it is not just the antioxidants and reduction in inflammation that account for the health outcomes, but the fact that olive oil may switch off genes that cause inflammation in arteries and blood vessels[306].

'Bad fats' message led to bad foods – the sugar curse

Some years ago, research on saturated fats led to the development of a host of low-calorie foods. The trouble is that they led to shelves filled with nutrient-light products, artfully filled with sugar, refined carbohydrates and additives to compensate for the loss of taste when fat is removed or lowered.

An astonishing 74% of packaged foods contain sugar! There is sugar in bread, soups, sauces, ketchup, sausages, tinned beans, lasagne, flavoured yogurt, breakfast cereals. And that's on top of obvious sources like sweets, jams and soft drinks.

It is all made worse by the fact that manufacturers cunningly list forms of sugar separately on labels. In fact, there are 61 different names for sugar listed on food labels! They vary from fructose, sucrose, glucose, agave, corn syrup, maltose, dextrose, barley malt, rice syrup, lactose and, when they are being honest, sugar.

As a result, we now eat about 20 times more sugar than 100 years ago. As we saw earlier, average adult sugar consumption is a massive 26 kg / 57 lb in the USA, and 21 kg / 47 lb in the UK.

If you track the rise in obesity against the rise in low-fat (but high sugar) foods, you will see that the trend lines are almost exactly the same.

Several leading neurologists have made the same observation about the rise in sugar consumption and parallel rise in Alzheimer's. A 2016 animal study shows that high sugar levels cause inflammation in the brain and inflammation is strongly linked to Alzheimer's[307].

Between 1970 and 2005, the average American and Briton unwittingly added 4 teaspoons a day of sugar to their diet, which is 76 calories. It doesn't sound a lot, but all things being equal, over a year that could add 8 lb/4kg of body weight. Annually! Worse still, excess sugar seems to preferentially increase visceral (abdominal) fat – the most dangerous type – and 'apple obesity'.

As Rafael Perez-Escamilla, an epidemiology professor at the Yale School of Public Health, points out:

> *"If we compensate for lower fat by increasing higher sugar or carbohydrates, all that does is increase the risk for obesity, diabetes, coronary heart disease".*

There is a direct connection between sugar and increased risk of heart disease. Sugar inhibits an enzyme that breaks down triglycerides (fats in the blood) and lowers the level of the protective HDL cholesterol.

What is *really* stupid is that the taxpayer subsidises the production of corn sugar/syrup in the USA and beet sugar in Europe. And then pays for the cost of diabetes and Alzheimer's treatment through healthcare.

The facts about our opening contradictions

Coffee is good for you

By far the largest study on coffee was published in 2019 and covered 3.8 million people in the USA, Asia and Europe. It found the lowest risk of 'all-cause mortality' at a 'dose' of 3.5 cups a day, and the lowest risk of cardio-vascular disease at 2.5 cups a day – irrespective of age. There was no effect on cancer mortality. So, coffee at between 2-4 cups a day or less is safe and may be heart beneficial[308].

This appears slightly puzzling as caffeine does initially seem to have a short-term spike on blood pressure, but this effect seems to decline quite rapidly.

Eggs are healthy

Eggs are an excellent source of inexpensive and high-quality protein. Over half the protein from an egg – about 6g – is found in the egg white. Eggs are also good sources of selenium, B6, B12, vitamin D, lutein and zeaxanthin, and minerals such as zinc, iron and copper.

We have already seen that the body makes its own cholesterol in the liver and adjusts to increased dietary cholesterol by making less internally. So concerns that eating eggs elevates cholesterol levels have now been shown to be wrong – in fact, studies show that eating an average of one to two eggs a day can actually increase HDL (the good) cholesterol.

For pre-diabetics with impaired glucose tolerance or actual diabetics, a study published in the *British Journal of Nutrition* concluded that a *"high-protein, energy-restricted diet, high in cholesterol from eggs, improved glycaemic and lipid profiles, blood pressure and apo-B in individuals with type-2 diabetes"*. Apo-B is the main protein in LDL, the 'bad' cholesterol.

There is no evidence that people who eat an average of an egg per day have higher rates of heart attacks, strokes or other cardiovascular diseases.

Butter is good...ish, but in moderation

Butter is 63% saturated fat, 26% monounsaturated fat and 4% polyunsaturated fat. It is also a source of butyrate which, as we have seen, is gut healthy and anti-inflammatory.

Two recent meta analyses, funded amongst others by British Heart Foundation, Cambridge National Institute for Health Research and the Canadian Institutes of Health Research, have found no proven statistical association between consumption of saturated fats and raised mortality or heart disease risk[309].

Nevertheless, not proven is not the same as no link, and butter is high in calories and best enjoyed in moderation alongside fats which are not just neutral in respect of heart disease, but positively heart healthy. Such as olive oil and omega-3 oils.

Butter can be useful for high-heat cooking, as it is resistant to oxidation and has a high 'smoke point'. This can help prevent the build-up of harmful free radicals when cooking. Avocado oil has the same benefits.

Alcohol – the jury is confused

Alcohol works fast. Thirty seconds after your first swallow, alcohol speeds into your brain. Neurotransmitters work less fast, slowing your reflexes and altering your mood.

There is no doubt that excess alcohol (over 3 units a day) is bad for you – but what about moderate social drinking, especially wine?

Results from an innovative study were reported in 2015. Some 224 diabetic patients, who did not normally drink, were randomly assigned to 150 ml (5 fl oz) of mineral water, white wine, or red wine with dinner for two years. All groups followed a Mediterranean diet without calorie restriction. After two years, no material differences were identified across the three groups in blood pressure, adiposity (body fat), liver function, or quality of life[310].

The conclusion:

> *"Moderate wine intake, especially red wine, among well-controlled diabetics as part of a healthy diet is apparently safe and modestly decreases cardiometabolic risk".*

Of course, these were diabetic patients. But is moderate alcohol good for non-diabetics? Unfortunately, the results are not consistent.

A review in the journal *Circulation* states that *"most evidence suggests that low-to-moderate red wine consumption is good for the heart"* [311].

On the other hand, a meta-study published in the *Lancet* concluded that even very small amounts of alcohol raise a drinker's risk for cancer and early death.

> *"We found that the risk of all-cause mortality, and of cancers specifically, rises with increasing levels of consumption, and the level of consumption that minimises health loss is zero"* [312].

A very large study in China on 512,000 adults over 4 years reported that:

> *"Alcohol consumption uniformly increases blood pressure and stroke risk, and appears in this one study to have little net effect on the risk of myocardial infarction (heart attack risk)."*

However, the trial was mostly on men and most drank spirits, not wine. What the study did hint at, however, is that variations in how alcohol affects genes may explain the conflicting results from so many studies.

Personally, I wouldn't like to contemplate a life without an occasional glass of red wine, but I am conscious that alcohol has all these effects:

- **You won't sleep as well.** Your body processes alcohol throughout the night and once the effects wear off, you do not get the good REM sleep your body needs to feel restored. You will probably also wake up to make a trip to the bathroom.

- **Alcohol affects your immune system.** Your body does not make quite as many white blood cells as normal to fight germs.

- **Alcohol interferes with how the pancreas makes insulin**. In heavy drinkers, alcohol can damage both the pancreas and liver, which can lead to fatty liver and diabetes.

- **Alcohol can cause DNA damage, which is why studies consistently show an increased risk for cancers.** The increase is minor in light drinkers, but significant as consumption increases.

- **Too much alcohol over time can shrink your brain**, affecting your ability to think, learn, and remember things.

If I were the judge on an alcohol jury, I would say keep off spirits, and if you want to drink at all, limit yourself to a glass (women) or maximum two (men) of mostly red wine and certainly do not drink every day.

Why the emphasis on red wine? Because the skins are kept in the vats when grapes are pressed to make red wine, but the skins are removed when making white wine. Since so many of the beneficial polyphenols are in the skin, red wine has ten times as many polyphenols as white.

I also think the balance is slightly in favour of moderate wine as opposed to no wine, because there is some evidence that our gut microbes not only like dark chocolate, they 'like' red wine too!

A small Spanish study[313] showed that red wine consumption can positively change gut microbe profile which *"suggests possible prebiotic benefits associated with the inclusion of red wine polyphenols in the diet"*.

A final point. It is true that red wine does contain resveratrol, which is claimed to have a long list of health benefits. But you can also get resveratrol from grapes, grape seeds, blueberries, dark chocolate, pistachios and peanuts.

Besides, scientists at the Johns Hopkins University School of Medicine have pointed out that the amount of resveratrol in the animal experiments which suggested resveratrol's possible benefits, was far higher than you could consume in food or wine or even in most supplements.

Vitamin supplements – it depends on what's in them

Some 38% of adults in the UK take a multivitamin supplement and the percentage is over 50% in the USA. Some doctors support them as a health 'insurance' and some dismiss them as a waste of money that just creates 'expensive urine'.

Who is right?

Doctors are not always the best source of up-to-date advice, as the typical time spent on nutrition lectures in a five-year medical degree course is just a few hours. Indeed, Daniel Lieberman of Harvard University is on record saying that:

"It is possible for a doctor to go through medical school without any nutritional training at all".

The most comprehensive review of doctors' nutrition training, published in 2019 in the *Lancet*, concluded:

"Nutrition is insufficiently incorporated into medical education, regardless of country, setting, or year of medical education.

"Deficits in nutrition education affect students' knowledge, skills, and confidence to implement nutrition care into patient care" [314].

So where did the suggestion come from that people in high income countries get all the nutrition they need from food and should stop wasting their money on supplements?

One credible source was a 2013 editorial in the US *Annals of Internal Medicine*, which got significant media coverage. But the media did not pick up on a subsequent rebuttal by Professor Meir Stamfer of Harvard Medical School who was amazed that *"such a poorly done paper should have been published in a prominent journal"*. (A quote I picked up in Bill Bryson's highly entertaining book 'The Body'.)

So, instead of only asking the doctors, let's look at the evidence.

Research study results

The **Physicians' Health Study II** is one of two large studies completed on the effects of a multivitamin/mineral supplement. A large group of male physicians – 22,000, all over the age of 55 – took either a multivitamin or a placebo pill for more than a decade[315].

The results have been mixed, with modest reductions in cancer and cataracts, but no protective effect against cardiovascular disease or declining mental function. This finding was supported by another study published in 2013[316].

In the **French study** 'Supplémentation en Vitamines et Minéraux Antioxydants', researchers randomly assigned 13,017 adults to receive a placebo or a daily supplement containing modest amounts of vitamin C (120 mg), vitamin E (30 mg), beta-carotene (6 mg), selenium (100 mcg), and zinc (20 mg).

After 7 years of use, the supplements lowered total cancer incidence and all-cause mortality in men but not women. But again, the supplements provided no protection against cardiovascular disease.

In the **Age-Related Eye Disease Study** (AREDS), all 3,640 subjects had varying degrees of AMD – age-related macular degeneration. They were randomly assigned to receive a placebo or a daily supplement containing high doses of vitamin C (500 mg), vitamin E (400 international units [IU]), beta-carotene (15 mg), zinc (80 mg), and copper (2 mg).

Over an average follow-up period of 6.3 years, the supplements significantly reduced the risk of developing advanced age-related macular degeneration and reduced loss of visual acuity[317].

Although these and other studies have shown that taking a multivitamin is safe, apart from the very positive AREDS eye study, they don't exactly represent a ringing endorsement.

However, if you have been following the evidence on foods and nutrients in the preceding chapters, you should not be surprised.

Protecting against heart disease, cancer and dementia and extending your health-span takes far more than a simple vitamin pill – or even vitamins plus minerals. As we have already argued, it should include nutrients like polyphenols, carotenoids, omega-3, betaine and CoQ10.

A typical vitamin and mineral pill provides only a simple baseline and it is formulated to a price. I have been present at a meeting with a team from a very well-known international vitamin and mineral brand. When invited to include a polyphenol blend into the product, they declined, saying: *"It may be better, but it would mean a price rise and we need to keep XX as cheap as possible"*.

Nutritional values of food <u>have</u> declined

Donald Davis and researchers from the University of Texas published a study on nutrient values for 43 different vegetables and fruits in the *Journal of the American College of Nutrition*. It covered the 50-year period from 1950 to 1999. They found *"reliable declines"* in the amounts of protein, calcium, phosphorus, iron, riboflavin (vitamin B2) and vitamin C over this period.

Davis and his colleagues explained the decline by agricultural practices that concentrated on improved size, growth rate, and pest resistance at the expense of nutrition. The report commented:

> *"Efforts to breed new varieties of crops that provide greater yield, pest resistance and climate adaptability have allowed crops to grow bigger and more rapidly, but their ability to manufacture or uptake nutrients has not kept pace with their rapid growth."*

Combining studies reported by the Kushi Institute, the Worldwatch Institute, the British Food Journal and a recent report in the New Scientist, it appears that there have been nutrient declines in food as follows:

Vitamin A ↓21%   Vitamin B2 ↓38%   Vitamin C ↓30%

Iron ↓37%   Magnesium ↓30%   Zinc ↓28%

There were also some reductions for vitamins B6 and vitamin E.

The trend applies also to animal products. Meat, eggs and dairy products from animals raised on pasture as opposed to intensive farming are higher in vitamins A, D, E and beta-carotene, and have a better ratio of omega-3 to omega-6 fatty acids.

However, not every scientist agrees that nutritional values have declined, and a 2017 report in *Science Direct* questioned the evidence[318].

What is undeniably true, however, is that plant breeders have tried to make their crops more palatable – for example, increasing the sweetness of carrots and reducing the bitterness of kale. But in doing so they have sacrificed some nutritional value.

The problem is made worse since the increasing use of chemical pesticides can reduce the uptake of minerals by plants. Moreover, food which used to be consumed near to where it was grown and did not have to travel far, is now transported vast distances. Fruits are often picked before they are ripe (ie. before they reach their full nutrient value) to increase their storage time.

Conclusion? It is probable that nutrient values in foods are lower than 50 years ago.

Dairy is anti-inflammatory, particularly for diabetics

We have seen again and again that inflammation is a driver of ageing and age-related illness. And some reports have said that dairy is inflammatory because of the relatively high levels of saturated fat. But is it?

A 2017 systematic review of 52 previous clinical trials investigated the impact of the consumption of dairy products on several dozen known inflammatory markers – did the intake of milk products create inflammation in the body?

The meta-survey clearly indicated an <u>anti</u>-inflammatory activity in humans (other than those who were allergic to bovine milk).

> *"Our review suggests that dairy products, in particular fermented products, have anti-inflammatory properties in humans not suffering from allergy to milk, in particular in subjects with metabolic disorders"* [319].

The study was not financed by the milk industry.

Their final comment is interesting – that dairy appears to have a particularly positive anti-inflammatory effect on people with metabolic disorders, including diabetics.

If you are lactose intolerant, you are likely to be aware of the problem and avoid dairy. But for most people, dairy – especially lower fat dairy and fermented milk products like yogurt and "real" cheese – can be considered as a positive part of your diet.

The American Institute of Cancer Research World monitors studies on a continuous basis and states (2020) that there is: *"suggestive evidence that dairy products may decrease the risk of colorectal cancer and breast cancer in pre-menopausal women cancer"*.

Dairy is a good source of vitamin D, calcium, magnesium and protein. Of course, so are other foods, so while dairy may be beneficial, for most people it is not essential.

And cheese? Cheese is a fermented milk product and traditional cheeses – like Stilton, Mozzarella, Cheddar, Gruyere, Emmental, Roquefort, Gorgonzola, and Camembert – contain friendly microbes. So, these may improve the diversity of your microbiome, which we have seen is an important health marker. However, since cheese is animal fat, it is wise to eat it in moderation.

Note, too, that processed cheeses (like cheese slices, soft cheese triangles, roundels in red wax) are a mixture of a small amount of real cheese with other high-fat dairy products for flavour and texture, and they have very little nutritional value.

The advice from Dr Simin Meydani, at the Nutritional Immunology Laboratory at Tufts University seems a sensible conclusion:

> *"Overeating full-fat dairy or sugar-sweetened dairy can contribute to weight gain – and obesity itself is associated with chronic inflammation. Controlling weight is important in terms of reducing inflammation, sticking to low-fat dairy choices can help control weight and help reduce inflammation"*.

Occasional natural meat is not harmful – but avoid processed meat

Unless you are a vegetarian or vegan for ethical reasons, the evidence is that occasional natural non-processed meat is not harmful to health, and has benefits.

Dr Frank Hu, Chair of the Department of Nutrition at Harvard Medical School, says in a 2020 article that:

"An accumulated body of evidence shows a clear link between high intake of red and processed meats and a higher risk for heart disease, cancer, diabetes, and premature death.

"The evidence shows that people with a relatively low intake have lower health risks. A general recommendation is that people should stick to no more than two to three servings per week. Consider red meat a luxury and not a staple food."

As for processed meat, the American Institute for Cancer Research confirms there is a much stronger association with a higher risk of heart disease and cancer, especially colorectal and prostate cancer.

The key words in Dr Hu's statement are 'high intake'. I suggest that 'high' for red meat – beef, pork and lamb – would be more than one serving a week. That is also the recommendation of the MIND diet, a diet developed to cut the risk of dementia. As for processed meat – like sausages, ham, bacon, salami and smoked meats – it is safer to simply avoid them.

There is nothing in red meat that you cannot get from poultry, fish, eggs and nuts, or by following a carefully planned plant-based diet.

The other factor is that meat and fish are frequently cooked using very high temperatures during frying, grilling or barbecuing (charbroiling). This creates compounds called heterocyclic amines (HCAs) and polycyclic aromatic hydrocarbons (PAHs). Both these have been linked to colorectal cancer.

This risk is increased when meat is cooked with intense heat over a direct flame, resulting in fat dropping on the hot fire, causing flames. These flames contain the harmful PAHs that stick to the surface of food. So 'flame-grilled' should probably be seen as a warning, rather than a great advertising slogan!

If you do BBQ, then marinating with herbs appears to lead to less HCAs.

Meat, milk and eggs are the main natural food sources of vitamin B12, which is vital for keeping the body's nerve and blood cells healthy and to help in the production of DNA.

A deficiency in B12 will manifest as tiredness. B12 is not generally present in plant foods, but vegans can obtain it from nutritional yeast and fortified cereals – or a supplement.

The HDL cholesterol debate – what is optimum?

The British Heart Foundation states that: *"In general the lower the LDL and non-HDL, the better, and higher the HDL, the better."*

Your blood cholesterol is measured in mg/dl (mg per decilitre) or mmol/L, which stands for millimoles per litre. The generally recognised desirable level of HDL cholesterol is above 40 mg/dl for men and 50 mg/dl for women – ie. above 1.0 and 1.2 mmol/L.

Although researchers at the Albert Einstein College of Medicine have suggested that a high HDL level is a marker of longevity, other researchers are not so sure. A 2016 study at Washington University School of Medicine concluded that it is *"intermediate HDL cholesterol levels that may increase longevity."* In support, a 2018 report from the European Society of Cardiology, based on 6,000 individuals, indicates that HDL cholesterol of between 41-60 mg/dl (1.1-1.5 mmol/L) had the lowest risk of heart attack or cardiovascular death.

The website acceleratedlearning.com/delay-ageing has a printable list of age measurements – including HDL, LDL and total cholesterol markers – that you can ask for in a blood test.

- 20 -

Climate crisis, loss of biodiversity and the ethics of food

Our food choices shape our world.

This book is about delaying ageing, in order to postpone or even prevent age related disease. But it makes no sense to live longer and healthier in a sicker world.

Although health research points to the need to eat a mainly a plant-based diet, the evidence also shows that the inclusion of fish and occasional meat can be part of a healthy longevity diet.

For those who are vegetarian or vegan – which includes my two younger children – their food choices are already planet friendly. But what if you see no health reason, as opposed to animal welfare or planet-friendly reasons, not to remain an omnivore?

My suggestions are:

Choose organic fruit and vegetables, particularly for the *Dirty Dozen*™.

Pesticide residues are not the only reason to consider organic produce. Other positives include the fact that conventional, high intensity farming is having an increasingly detrimental effect on the environment and that organic crops have generally higher nutritional values.

In 2014, the UK's Newcastle University reviewed 343 studies on produce and found pesticide residues to be four times higher in conventional crops than in organic. The same study concluded that organic crops contained significantly higher antioxidant levels than their non-organic counterparts.

Why? Plants that are not sprayed with pesticides develop stronger defences, hence their higher protective antioxidant levels.

As we have seen, this protection is mostly in the form of flavonoids and polyphenols. So much so, that the University of Newcastle study estimates a switch to organic fruits, vegetables and cereals could provide additional antioxidants equivalent to eating between 1-2 extra portions of fruit and vegetables a day.

Cut down on meat. It is an astonishing fact that over 70% of all agricultural land is either used for raising animals for meat or dairy – or used for growing the feed for those same animals[320].

If we simply cut our meat consumption by half, or ideally two thirds, we would be healthier and so would the planet. It is now so much easier to do this when there are very many tasty and easily available vegetarian and vegan alternatives to meat.

Meat and dairy averagely provide just 18% of our calories and 37% of our protein but produce 60% of agriculture's greenhouse gas emissions[321].

Omnivores, which include me, cannot duck the conclusion that it is generally better to cut out the animal middleman and get most of your protein direct from plants.

Cut down on cow's milk for the same reason. The plethora of available milk alternatives now makes it so much easier. I often mix cow's milk and soy, oat, hemp or pea protein milk for cereal. It's an easy way to cut consumption in half.

Choose meats that have the least adverse impact on the environment. Beef is the worst, as it typically takes 7-8 kg of feed to create 1 kg of beef. That is followed by pork – with a feed to meat ratio of 3:1. Lamb is next, and free-range chicken at a 2:1 ratio probably creates the least environmental damage or animal welfare issues.

However, I can see the vegans' argument that you can hardly talk about animal welfare when the end-product is a dead animal. And intensive battery chicken rearing should be made illegal – if not just for animal welfare but to prevent a pandemic far, far worse than COVID-19.

This is not scaremongering. As long ago as 2005, the UN warned that CAFOs – concentrated animal feeding operations or factory farms – and wildlife 'wet' markets threatened to trigger a pandemic. There are Chinese and US chicken mega-farms housing literally multi-millions of chickens in tight, close, unsanitary contact that are potential breeding grounds for disease.

If the H7N9 virus ever mutated to become easily transmissible between humans, the US Centers for Disease Control estimate the mortality rate could be between 7,000 and 10,000 per 100,000 infected.

COVID-19 has a mortality rate of about 500 per 100,000.

In 2017, Robert Vries and colleagues of the Scripps Research Institute calculated that just three mutations would switch H7N9 to become human to human transmissible. It's a risk to our health and our economies that we really should not permit.

Organic meats are the gold standard – more sustainable with lower environmental impact – and not routinely treated with antibiotics.

Farming fish also has an adverse environmental impact, but the feed/protein ratios (typically 1:1) are more sustainable and there is a welcome trend for huge netted farms far out in the open ocean that are much less polluting.

Pay more for your food if you can possibly afford it. Cheap food comes at a high price. The industrialisation of agriculture in most of the West has changed farming from small, diversified, independently operated family farms into huge, specialised, agribusinesses growing largely a single monocrop. This brings big risks.

Food productivity has increased but is being bought at the long-term cost of soil erosion, contamination with agricultural chemicals, fertilisers, pesticides and a decline in pollinating insects. As a result, the real cost of food is not reflected on the supermarket shelf – it is merely deferred to future generations.

So, what can you do?

- Reduce animal protein. Buy organic if you can.

- Where you can, grow some of your own fruits, vegetables and mushrooms. This is possible even in an area as small as 2m x 2m or a window-box.

- For the fruits and vegetables you cannot grow, try to buy more locally and eat more seasonally, which cuts down on airmiles. Buy organic if you can.

- Resist supermarket multi-buys that encourage food waste. You can buy cheaply in high season and freeze.

- Boycott the cheap high fat/high sugar ready-made foods, which the manufacturers know full well are addictive. Their responsibility is to their shareholders, not your health.

Your cost will be more in money, but less in health, and the changes will be positive for the world and its wildlife.

People have made extraordinary, difficult and costly changes in lifestyle when faced with the immediate threat of the COVID-19 pandemic. Whilst the threat from climate heating is less immediately threatening, it is real.

So, is it too much to hope that this experience will make people and societies see that big changes for a better future can be made – if we have the will?

As Gandhi said:

"You must be the change you want to see."

- 21 -

A brave new world of healthcare?

As a finale, I thought you might like a glimpse of just some of the current developments in healthcare – some rather fun or even bizarre, some very significant.

Products and devices in development

A bathroom mirror that you breathe on when you get up and analyses your breath for oxygen levels and early signs of illness. Then it analyses your face for clues about your mental state[322]. But maybe rather a shock as a start to the day?

Bathroom scales that instantly tell you, not just your BMI, but your body fat percentage.

A toilet seat that alerts the user – and his or her health provider by automated text – of their risk of congestive heart failure[323].

Another toilet that analyses human waste to report on microbiome imbalances, urinary tract infections, and indicators of diabetes, infections, kidney disease and cancer[324].

Healthcare voice applications. Dr Alexa will hear you now? A version of Alexa (or Cordana or Siri) that analyses patterns in your voice for stress or even early signs of cognitive decline. No wonder Google wanted to buy Fitbit. There are over 37 start-ups developing healthcare voice applications.

A wearable chest strap that records and then communicates an electrocardiogram result (ECG) directly to your doctor[325].

Wearable health devices that monitor not just heart rate, blood pressure, respiration rate, but blood oxygen saturation, blood glucose, skin perspiration, and body temperature[325]. Even a wearable that monitors your alcohol level!

A bluetooth toothbrush with sensors that monitor your brushing technique and results and can send the data to your smartphone or even your dentist[326].

A pendant that alerts a patient's carer of a fall.

Nutrition smartphone apps that can scan any food and give you its nutritional value.

The smart fridges that will alert you when food is going off, automatically order anything that is running low, and even suggest a nutritionally balanced meal based on what's inside – one that will be compatible with your levels of activity and weight-loss goals[322].

Brown fat transplants. A company is planning to offer to transplant brown fat cells. Brown fat cells – which we have far more of when we are young – burn blood fats and sugars rather than storing them. By transplanting brown fat cells, the theory is that your metabolism will increase, and weight will reduce.

Stem cell transplants. The same company (AgeX Technologies), founded by Dr Michael West, whose team first discovered telomerase, is working to target human ageing and age-related degenerative diseases using pluripotent stem cells – stem cells that are self-replicating and can become any type of cell[327].

Quantum physics and health

Expect quantum technology to begin to have a big impact on healthcare over the next decade, because many cell processes take place at the nanoscale – atoms and subatomic particles.

For example, using a method known as the bio-barcode assay, scientists can now detect biomarkers in the blood using harmless gold nanoparticles. These are visible using MRI technology and have unique quantum properties that allow them to attach to cells that fight disease.

Researchers at the University of York and elsewhere are experimenting with these gold nanoparticles which are coated in antigens (substances that bind to antibodies) before being introduced to the body, where they are designed to attach to cancer cells.

The patient is then treated in an MRI machine that triggers the particles to heat up and destroy the cancer cells. When the machine is turned off, the particles cool back down and can be removed from the body without harm to the patient.

Although this technology is at relatively early stage, it may help solve the toxicity problem of current chemotherapy. Conventional cancer drugs do not differentiate well between cancerous and normal cells, leading to distressing side effects. This greatly limits the maximum dose of the drug that can be delivered.

Moving from sickcare to healthcare

All the above developments are either here now – or are a very few years away. Although I included a few zany ones for light relief, they herald a new, serious direction. From what is mostly currently a reactive 'sickcare' service, to a proactive healthcare service.

What have, until now, been dedicated monitoring devices at hospitals and doctors' surgeries are migrating to your smartphone. Apps are creating data designed to alert the patient and their healthcare professionals early enough to take preventative action, which is positive.

Monitoring and drug delivery via your smartphone

Wearable activity trackers, including smartphones, can already track your heart rate, movement, steps and sleep. They can, if they have an internal gyroscope, sense if you have been sitting too long and automatically send you a message to stand up and move around. They will measure your blood pressure and soon certain key biochemicals.

For pre-diabetic patients, that is a breakthrough, because high blood pressure, elevated cholesterol and excess body fat are linked together in a matrix.

For some others, I worry about a new cohort of 'cyberchondriacs'!

Currently, these wearables only communicate the results to you. But the next step will be, if you give them permission, to send that data direct to your doctor or even to your health club. But that raises real issues of data protection and data safety – which leaders in the field believe can be solved using block chain technology. (No, I don't really understand it either!).

There will also be wearable drug delivery systems to enable the doctor to authorise a remote adjustment to a drug dose depending on the data feedback from that device. They could also know in real time if the patient has not taken his meds, or is not exercising.

Personalised and preventive healthcare

While some of this may sound like shades of Big Brother, it does move the central figure in healthcare from the doctor to the patient. From treating the disease with standard 'commonly prescribed' drugs, to treating the whole individual patient with a more personalised regime.

Currently, a treatment delivered to two different patients can have two very different outcomes, depending on their genetic differences. In future, artificial intelligence (AI) will be deployed to scan a patient's genome and draw on vast databases to identify the type of drug that will best suit the patient.

We will move from a situation where medicine is practised at the point where the disease has surfaced, to a prior point where it can be predicted and hopefully prevented. That's the goal of the UK based Biogerontology Research Foundation[328], which is backing an AI Centre for Longevity based at King's College London.

An international team in 2019 announced that the tell-tale DNA mutations that are a hallmark of cancer could be detected many years before the disease surfaced. Which gives us the potential to treat early[329]. Earlier is always better in cancer. Application of this 'liquid biopsy' technique, however, is some way off, as another paper in 2020 cautions.

Individualised precision medicine

These developments will ultimately usher in the era of personalised, individualised, or precision medicine, where drugs and treatments are tailored to your specific genetic profile.

A patient's genomic data will be integrated into their medical record. Familial predispositions to a disease can alert caregivers to appropriate tests and what drugs may need to be avoided.

Personalised or 'precision' medicine is beginning to happen already in some cancer immune therapies. Cells can be removed from a patient and sent to a lab, where they are modified to attack specific cancer cells. These modified cells are then returned to the patient they were originally extracted from, where they attack and kill that patient's cancerous cells[330].

Building the 'All of Us' health database

Perhaps the most ambitious programme of all in the connected health field is the *All of Us* research programme funded by the US National Institutes of Health (NIH). It aims to recruit 1 million Americans from all backgrounds, to build one of the most diverse health databases in history.

Researchers will use the data to learn how our biology, lifestyle, and environment affect health. Their web site, if you are American and want to join, is: allofus.nih.gov.

All of Us will ultimately combine genomics, streaming data from wearables, bio-sample collections and medical records into a gigantic database. That database will be searched by AI for conclusions on the interaction between genotype, the set of genes in your DNA and phenotype, the physical expression of those genes as a result of your environment, your diet, activity levels, stress levels, etc.

The economic payoff?

There is an economic imperative to move to preventative healthcare and to devote more research money to the processes underlying ageing. Dr Jay Olshansky, board director of the American Federation for Aging Research, has stated that to delay ageing is *"a public health initiative on a par with the introduction of antibiotics"*, and has a potential saving in healthcare costs in the trillions.

In 1960, healthcare expenditure in the USA was 5% of GDP. By 2016, it had ballooned to 17.5%. In the UK that same year, it was 9.8%, counting both NHS and private spending.

This is a significant financial problem, when the demographics show that fewer working people will be supporting more retired people.

Giving doctors and nurses more human interaction time

Yet, despite these colossal expenditures, we are short of doctors and nurses in both the UK and USA, so they are stressed and overworked. A visit to a doctor usually lasts only a few minutes and a lot of that time the doctor is looking at a screen rather than the patient. As a result, errors can happen, and indeed misdiagnosis is the fourth-leading cause of death in the United States, trailing only heart disease, cancer, and stroke.

So, several teams are using voice to text software to develop systems that will take automatic notes during a doctor/patient consultation – which will free up the doctor to spend more face time with the patient.

Using AI to improve diagnosis

Another positive is the use of AI to potentially cut healthcare costs. Computers have been trained with millions of medical image scans and, using what is called 'Deep Learning', are enhancing radiology. This is already improving the accuracy of diagnosing breast, head and lung scans. Eric Topol gives an example in his state-of-the-art book 'Deep Medicine':

> *"There are over 20 million medical scans performed in the US every day. An MRI, for example, costs hundreds to thousands of dollars. AI could process 260 million medical scans (more than 2 weeks' worth) in less than 24 hours for a cost of only $1000. We pay billions and billions of dollars for the same work today."*

A reality check

This has been a very short romp through just some elements of the Digital Health Revolution, and we have not even mentioned tele-medicine, 3D-printed body parts, health robots, or tissue regeneration therapies that could regrow body parts like a kidney.

All this will necessitate enormous databases, a vast collection of personal data, massive computing power and the application of AI to spot correlations and trends. Which is why Apple, Google, Amazon, Microsoft, Kaiser Permanente, Accenture, 23andMe, Walgreen-Boots and hundreds of start-ups are all so heavily invested in it.

What is positive about these 'connected health' possibilities is that we will be much better informed, earlier, about how to manage our health and able to be more proactive participants in it. That is undeniably good.

But I want to introduce a reality check.

High-tech, low-tech or mid-tech?

Dr Joe Kvedar runs Partners Connected Health, an organisation whose mission is to *"leverage information technology – mobile phones, tablets, wearables, sensors and remote health monitoring tools – to deliver quality patient care outside of traditional medical settings"*.

So, Dr Kvedar is very much an evangelist for the new age of Digital Medicine. Yet, this is what he says in his book 'The New Mobile Age – How Technology Will Extend Healthspan':

> *"I have observed three predictors of longevity in my patients: companionship, a sense of purpose and moderate activity."*

Nothing very high tech there. And nothing that you don't get in Blue Zones.

Caveat 1 – increased inequality?

I feel uneasy about two potential issues with the new Digital Health developments. Firstly, that we may be heading for a situation where inequality could worsen in health. We already know that the higher your income, the better your health and life expectancy.

Alphabet (the umbrella company of Google) has already invested well over $1 billion into a company called Calico, with the aim of extending human lifespans. A lot of that money seems to be going into gene therapies, which have been shown to double life expectancy in mice, when two genes thought to be connected to ageing were knocked out.

Although gene therapy has the potential to cure some devastating genetic diseases, it is neither uncontroversial, nor cheap. *Luxturna* is the first true gene therapy product approved in the United States and treats – actually cures – an inherited form of blindness. The cost is $425,000 per eye.

But when gene therapy is contemplated as a way to increase lifespan, will we end up with a tiny group of super-rich, super-centenarians?

Caveat 2 – the assumption that we still get ill as we age

My second concern is that most of the new developments still assume that you will get ill, albeit they aim to pick up the signs earlier. Whereas the underlying concept behind the low-tech, inexpensive ideas in this book is that our concentration should be on preventing illness in the first place.

The *Delay Ageing Plan* in this book is affordable by most and is low tech. All it involves is a focus on nutrient dense foods – with a supplement to reach towards Blue Zone nutritional levels – a 37 minute a day activity plan, a stress reduction exercise, some mindfulness and a suggestion that you become more aware of what makes you happy and do more of it.

Even our suggestion that you track your efforts on a paper-based Tracking Wheel is very low tech. But it is doable by anyone.

That doesn't mean that incorporating health monitoring into your life isn't a good idea. It is, because the more aware of your health metrics you are, the more responsibility you can take for your own health.

So, I suggest the future should be 'mid-tech'.

That conclusion is reinforced by a recent article on the website *Future Medicine* which concluded that:

> *"In the case of many common chronic degenerative diseases, our individual genome may therefore have limited impact: our nutrition, exercise patterns and many aspects of our environment (ie. our phenotype) can be just as important".*

And those we can control.

CONCLUSION

THE BEGINNING OF THE BEGINNING

I hope that this book has inspired you to create yourself as an individual 'Blue Zone'.

I'm now 80 and am fit and well and on no medication. I am hoping to reach my late 90s, even 100, but with the health of a 60-year-old. I know it is possible. For you too.

This may be the end of the book – but I hope and trust that it's the beginning of a healthier, longer life for you and your family.

Together, we've examined the role of food, activity, supplements, mediation, relaxation. They are all important. But none more important as maintaining a sense of wonder at the amazing world in which we live and maintaining an enquiring mind.

Albert Einstein said:

"Do not grow old, no matter how long you live. Never cease to stand like curious children before the great mystery into which we were born."

So, let's conclude with the image we started with. A single cell.

And marvel "like curious children" that all the thousands of processes we have discussed start in that tiny, tiny space.

Colin Rose, September 2020

Join me, please

You'll have seen that delaying ageing is all about creating a rejuvenating, forward looking lifestyle.

So I hope you'll join me and take advantage of the free resources on acceleratedlearning.com/delay-ageing.

They include innovative, tasty, mostly plant-based recipes, the printable food plan, guided meditation, core strengthening exercises and frequent updates on the science of healthy longevity. Plus information on tests that measure biological rather than chronological ageing, that you can ask your doctor for.

Your opinion and reviews

I really hope you have found *Delay Ageing* interesting and helpful. If you did, can I ask you to leave a review on Amazon? It is the best way for authors to gain exposure and help sales. Thank you in advance.

Ongoing updates

You can follow me on Facebook where I post articles on health.

facebook.com/Nutrishield/

Or register for updates on ageing research at:

acceleratedlearning.com/delay-ageing

Contact

You can reach me as an author at:

Accelerated Learning Systems Ltd
12 The Vale, Southern Road, Aylesbury, Bucks HP19 9EW, UK
Tel: +44 (0)1296 631177
Email: info@acceleratedlearning.com

- APPENDIX 1 -

Ageing Research Centres and Researchers

Although this book is for the layperson, I hope you will want to keep updated on this exciting field. So, I have listed some centres and individuals who are prominent in the field. This is certainly very far from a complete list of researchers in Gerontology. However, I have tried to highlight some of the key names, who are working on one or more aspects of healthy ageing mentioned in this book.

NOTE: Links and researcher titles and institutions correct at time of publication.

Some research centres and their websites

Ageing Research at King's College London
kcl.ac.uk/health/research/divisions/cross/ark/ageing-research-at-kings

American Federation for Aging Research
afar.org

Aston (Birmingham UK) Research Centre for Healthy Ageing
aston.ac.uk/lhs/research/centres-facilities/archa

The Biogerontology Research Foundation
bg-rf.org.uk

Buck Institute for Research on Aging, Novato, California
buckinstitute.org

Centre for Cognitive Ageing and Cognitive Epidemiology – University of Edinburgh
ccace.ed.ac.uk

European Institute for the Biology of Ageing
eriba.umcg.nl

Glenn Foundation for Medical Research (has centres at Harvard, MIT,
Princeton, Stanford, Salk and Michigan as well as the Buck Institute and Albert Einstein
College of Medicine)
glennfoundation.org

Institute of Ageing and Chronic Disease, Liverpool University
liverpool.ac.uk/ageing-and-chronic-disease

Institute of Healthy Ageing, University College London
ucl.ac.uk/biosciences/departments/genetics-evolution-and-
environment/institute-healthy-ageing-iha

Leibniz Institute on Aging (Fritz Lipmann Institute), Germany
leibniz-fli.de

National Institute on Aging (US) – nia.nih.gov

Sheffield University Healthy Lifespan Institute
sheffield.ac.uk/healthy-lifespan

University of Texas Health Science Center
uth.edu/

Individual researchers and their main fields of study

Bruce **Ames** – University of California, Berkeley.
Age related degenerative diseases.

Steven **Austad** – University of Texas Health Science Center.
Biology of ageing.

Nir **Barzilai** – Institute for Ageing Research at the Albert Einstein
College of Medicine. Protective genetics of longevity.

Berenice **Benayoun** – University of Southern California.
Epigenetic change and ageing.

Elizabeth **Blackburn** – University of California, San Francisco.
A pioneer in the field of telomeres and telomerase, Nobel Prize winner
for her work on telomerase.

Maria **Blasco** – Spanish National Cancer Center. DNA repair, telomeres, cancer, and ageing. One of the authors of *The Hallmarks of Aging*.

Alex **Bokov** – University of Texas Health Science Center. Ageing and life-extension.

David **Botstein** – Chief Scientific Officer of Calico. Understanding the biology that controls lifespan.

Dale **Bredesen** – Buck Institute. Neurodegeneration and neural cell death.

Judith **Campisi** – Buck Institute. Cellular senescence, cancer, and ageing.

Claudia **Cavadas** – University of Coimbra. Dietary manipulations of ageing, brain ageing and neuro-degeneration.

Richard **Cawthorn** – University of Utah. Genetics of longevity, telomeres.

Danica **Chen** – University of California, Berkeley. Stem cell ageing, caloric restriction.

George **Church** – Harvard Medical School. Reversal of ageing with gene therapy, creating species resistant to multiple viruses by gene editing. (Recreated the woolly mammoth!)

David **Clancy** – Lancaster University. Ageing in fruit flies.

Francis **Collins** – Director of the US National Institutes of Health. Major role in sequencing of the human genome, ageing and longevity.

Lynne **Cox** – University of Oxford. Replicative senescence and DNA damage.

Ana Maria **Cuervo** – Albert Einstein College of Medicine. Autophagy and neurodegeneration.

Sean **Curran** – University of Southern California. Ageing and genes.

Andrew **Dillin** – Salk Institute for Biological Studies. Development of ageing and age-related diseases primarily using *C. elegans*.

Ian **Dreary** – University of Edinburgh. Cognitive ageing.

Monica **Driscoll** – Rutgers University.
Mechanisms of ageing, neuronal function.

Richard **Faragher** – University of Brighton.
Cellular senescence and Werner's syndrome.

Caleb **Finch** – University of Southern California.
Brain ageing, and evolutionary biology of ageing.

Toren **Finkel** – US National Heart, Lung and Blood Institute.
Signalling pathways in ageing and disease.

Luigi **Fontana** – University of Sydney.
Longevity, calorie restriction and exercise.

Claudio **Franceschi** – University of Bologna.
Work on centenarians, inflammation and ageing.

Lazaros **Foukas** – University College London.
Cell signalling pathways.

George **Garinis** – Institute of Molecular Biology and Biotechnology
Greece. DNA damage and repair, progeroid syndromes.

David **Gems** – University College London.
Genetics and the evolution of ageing.

Jesus **Gil** – London Institute of Medical Science. Senescent cells.

Cristina **González-Estévez** – Leibniz Institute on Aging.
Calorie restriction.

Vera **Gorbunova** – University of Rochester, USA.
DNA repair, telomeres and telomerase in mammalian models.

Helen **Griffiths** – University of Surrey.
Inflammation, nutrition and ageing.

Beatrix **Grubeck-Loebenstein** – University of Innsbruck.
Immunology.

Leonard **Guarente** – Massachusetts Institute of Technology.
Cellular models of ageing, caloric restriction, sirtuins.

Tory **Hagen** – Oregon State University.
Free radicals and mitochondrial ageing.

Lorna **Harries** – University of Exeter.
Gene regulation and ageing.

Paul **Hasty** – University of Texas Health Science Center.
DNA repair and cell cycle response to DNA damage.

Leonard **Hayflick** – University of California, San Francisco.
Pioneer in cellular senescence.

Siegfried **Hekimi** – McGill University, Canada. Genetics of ageing.

Jan **Hoeijmakers** – Erasmus Medical Center, Netherlands.
DNA repair.

Peter **Hornsby** – University of Texas Health Science Center.
Cellular ageing and tissue engineering.

Steve **Horvath** – UCLA. Genomic biomarkers of ageing.

Malcom **Jackson** – University of Liverpool.
Muscle ageing, free radicals and nutrition.

Thomas **Johnson** – University of Colorado at Boulder.
Identified the gene (named *age-1*), able to delay ageing.

Leanne **Jones** – University of California, San Diego.
Stem cells and their role in cancer and ageing.

Jan **Karlseder** – Salk Institute, California.
Telomeres and DNA repair.

Brian **Kennedy** – Buck Institute.
Nutrient signalling pathways and dietary restriction.

Cynthia **Kenyon** – University of California, San Francisco.
A pioneer in the genetics of ageing.

Kerri **Kinghorn** – University College London.
Neuro-degenerative disorders.

David **Kipling** – Cardiff University.
Cellular senescence, telomeres and telomerase.

Guido **Kroemer** – Centre de Recherche des Cordeliers, Paris.
Intra- and extra-cellular stress pathways. One of the authors of *The Hallmarks of Aging*.

Makoto **Kuro-o** – University of Texas Southwestern Medical Center. The klotho gene.

Dudley **Lamming** – University of Wisconsin. mTOR and longevity.

Peter **Lansdorp** – University of Groningen, Netherlands. Director of the European Institute for the Biology of Ageing.
Genetic instability in ageing and cancer, stem cell biology.

Pamela **Larson** – University of Texas Health Science Center. Ageing research using *C. elegans.*

David **Le Couteur** – Director of the Centre for Education and Research on Ageing of the University of Sydney.
Nutrition and ageing.

Virginia **Lee** – University of Pennsylvania School of Medicine. Neurodegenerative diseases.

Christian **Leeuwenburgh** – University of Florida. Ageing, oxidative stress.

Rodney **Levine** – National Heart, Lung, and Blood Institute (USA). Oxidative damage and free radicals.

Gordon **Lithgow** – Buck Institute. Ageing research on *C. elegans.*

Valter **Longo** – University of Southern California. Mechanisms of ageing.

Carlos **Lopez-Otin** – Universidad de Oviedo, Spain. Proteolytic enzymes. One of the authors of *The Hallmarks of Aging.*

Janet **Lord** – University of Birmingham (UK). Immune system, inflammation, stress and ageing.

Vicki **Lundblad** – Salk Institute, California. Telomerase.

Michael **Lustgarten** – Tufts University, Massachusetts. Microbiome and ageing.

Alvaro **Macierira-Coelho** – French National Institute of Health. Cell senescence.

George M **Martin** – University of Washington, Seattle. Alzheimer's, biology of ageing.

Edward **Masaro** – University of Texas Health Science Center.
Caloric restriction.

Mark **Mattson** – National Institute on Aging (USA).
Neurodegenerative diseases.

Anne **McArdle** – University of Liverpool.
Muscle ageing, oxidative stress and nutrition.

Simon **Melov** – Buck Institute. mDNA damage.

David **Melzer** – Exeter University.
Epidemiology and genetics of longevity in humans.

Qing-Jun **Meng** – University of Manchester.
Biological clocks and ageing and disease.

Brian **Merry** – University of Liverpool.
Caloric restriction, oxidative stress in ageing.

Richard **Miller** – University of Michigan.
Genetics and biology of ageing in mice, the immune system.

Pedro **Moradas-Ferreira** – Institute for Molecular and Cellular Biology
Portugal. Oxidative stress.

Richard **Morimoto** – Northwestern University, Chicago.
Cellular quality control, chaperones and stress responses.

Florian **Muller** – Dana-Farber Cancer Institute, Boston.
Mitochondrial free radical production.

Coleen **Murphy** – Princeton University.
Mechanisms of ageing in *C. elegans*.

Ellen **Nollen** – University of Groningen, Netherlands.
Proteins, neurodegeneration and ageing.

Catarina **Oliveira** – University of Coimbra, Portugal. Brain ageing.

Tiago **Outeiro** – University of Newcastle. Neurodegenerative diseases.

Linda **Partridge** – University College London. Director of the Max
Planck Institute of the Biology of Ageing. Evolutionary theory of
ageing. One of the authors of *The Hallmarks of Aging*.

Joao **Passos** – Mayo Clinic, USA.
Cellular senescence, oxidative stress and mitochondria.

Chris **Patil** – Buck Institute. Life extension.

Olivia **Pereira-Smith** – University of Texas Health Science Center.
Cellular senescence, cancer and ageing.

Thomas **Perls** – Boston University. New England Centenarian Study,
genetic factors of ageing in centenarians.

Peter **Piper** – University of Sheffield. Senescent cells.

Iryna **Pishel** – Institute of Gerontology, Ukraine.
Ageing of the immune system.

Matthias **Platzer** – Leibniz Institute on Aging.
Genomics of cancer and ageing.

Paul **Potter** – Medical Research Council, Harwell, UK. Genetics of
ageing, mutations that impact ageing and age-related diseases.

Lawrie **Rajendran** – King's College London. Dementia research.

Thomas **Rando** – Stanford University.
Stem cells and muscle ageing.

Karl **Riabowol** – University of Calgary, Canada.
Cellular senescence, DNA repair and cancer.

David **Rollo** – McMaster University, Canada.
Ageing and nutrition.

Michael **Rose** – University of California at Irvine.
Evolution of ageing.

Corinna **Ross** – Texas Biomedical Research Institute.
Mammalian ageing and obesity.

Karl Lenhard **Rudolph** – Leibniz Institute for Age Research, Germany.
Telomeres, stem cells and ageing.

Anders **Sandberg** – Oxford University. Cognitive ageing.

Heidi **Scrable** – Mayo Clinic, USA.
Genetic manipulations in mice to understand ageing.

John **Sedivy** – Brown University, USA.
Cellular senescence and epigenetics.

Colin **Selman** – University of Aberdeen.
Mechanisms of ageing and caloric restriction.

Manuel **Serrano** – Institute for Research in Biomedicine, Barcelona.
Ageing and cancer. One of the authors of *The Hallmarks of Aging*.

Jerry **Shay** – University of Texas.
Cellular senescence, telomeres and telomerase.

Paul **Shiels** – University of Glasgow.
Biological ageing, telomeres and cell senescence.

Robert **Shmookler-Reis** – University of Arkansas.
Metabolism, DNA methylation, Werner's syndrome, genetics of ageing.

David **Sinclair** – Harvard University.
Caloric restriction, sirtuins and resveratrol.

Eline **Slagboom** – Leiden University Medical Center, Netherlands.
Epigenetics and metabolic disease

William **Sonntag** – University of Oklahoma Health Science Center.
Neuro-endocrine mechanisms of ageing.

John **Speakman** – University of Aberdeen.
Metabolism, energetics, obesity and ageing.

Tim **Spector** – King's College London. Twin studies related to genetics
of diseases, including age-related diseases.

Claire **Stewart** – John Moores University, Liverpool.
Stem cells, exercise and ageing.

Alexandra **Stolzing** – Loughborough University.
Stem cell biology and regenerative medicine.

Yousin **Suh** – Albert Einstein College of Medicine, New York.
Genetics of ageing and longevity.

Fushing **Tang** – University of Arkansas. Diet and ageing.

Paul **Thornalley** – Warwick University Medical School, UK.
Protein damage in diseases and ageing.

Janet **Thornton** – European Bioinformatics Institute, Cambridge, UK. Structural biology and bioinformatics.

Heidi **Tissenbaum** – University of Massachusetts. Molecular mechanisms of ageing.

John **Trojanowski** – University of Pennsylvania School of Medicine. Neurodegenerative diseases.

Jennifer **Tullet** – University of Kent. Longevity research with *C. elegans*.

Jan **Van Deursen** – (previously) Mayo Clinic. Senescent cells.

Gary **Van Zant** – University of Kentucky. DNA repair, stem cells, cancer and ageing.

Eric **Verdin** – Buck Institute. Ageing, epigenetics and the immune system

Jan **Vijg** – Albert Einstein College of Medicine, New York. DNA mutation and repair in ageing.

Manlio **Vinciguerra** – University College London. Metabolism, liver diseases and ageing.

Christi **Walter** – University of Texas Health Science Center. DNA repair, cancer and ageing.

David **Weinkove** – Durham University. Ageing and development in *C. elegans*.

Stephen **Welle** – University of Rochester, USA. Ageing, muscle metabolism and gene expression.

Rudi **Westendorp** – Leiden University Medical Center, Netherlands. Epidemiology and genetics.

Catherine **Wolkow** – National Institute on Aging, USA. Insulin-like pathways and the nervous system.

Woodring **Wright** – University of Texas Southwestern Medical Center. Cellular senescence, telomeres and telomerase.

Thomas von **Zglinicki** – Newcastle University. Cell senescence, mitochondria, telomeres and telomerase

- APPENDIX 2 -

References

1. López-Otín, C., Blasco, M. A., Partridge, L., Serrano, M. & Kroemer, G. The hallmarks of aging. *Cell* vol. 153 1194 (2013).

2. Campisi, J. *et al.* From discoveries in ageing research to therapeutics for healthy ageing. *Nature* vol. 571 183–192 (2019).

3. Baker, D. J. *et al.* Clearance of p16 Ink4a-positive senescent cells delays ageing-associated disorders. *Nature* **479**, 232–236 (2011).

4. Heart ageing - Press Office - Newcastle University. https://www.ncl.ac.uk/press/articles/archive/2019/02/heartageing/.

5. Bussian, T. J. *et al.* Clearance of senescent glial cells prevents tau-dependent pathology and cognitive decline. *Nature* **562**, 578–582 (2018).

6. Yousefzadeh, M. J. *et al.* Fisetin is a senotherapeutic that extends health and lifespan. *EBioMedicine* **36**, 18–28 (2018).

7. Madeo, F., Bauer, M. A., Carmona-Gutierrez, D. & Kroemer, G. Spermidine: a physiological autophagy inducer acting as an anti-aging vitamin in humans? *Autophagy* vol. 15 165–168 (2019).

8. Eisenberg, T. *et al.* Cardioprotection and lifespan extension by the natural polyamine spermidine. *Nat. Med.* **22**, 1428–1438 (2016).

9. Minois, N. Molecular Basis of the ‘Anti-Aging' Effect of Spermidine and Other Natural Polyamines - A Mini-Review. *Gerontology* **60**, 319–326 (2014).

10. Matsumoto, M. & Benno, Y. Consumption of Bifidobacterium lactis LKM512 yogurt reduces gut mutagenicity by increasing gut polyamine contents in healthy adult subjects. *Mutat. Res. - Fundam. Mol. Mech. Mutagen.* **568**, 147–153 (2004).

11. Schafer, M. J. *et al.* Exercise Prevents Diet-Induced Cellular Senescence in Adipose Tissue. *Diabetes* **65**, 1606–1615 (2016).

12. Huang, S. Inhibition of PI3K/Akt/mTOR signaling by natural products. *Anticancer. Agents Med. Chem.* **13**, 967–70 (2013).

13. Antunes, F. *et al.* Autophagy and intermittent fasting: the connection for cancer therapy? *Clinics* **73**, (2018).

14. Xu, M. *et al.* Senolytics improve physical function and increase lifespan in old age. *Nat. Med.* **24**, 1246–1256 (2018).

15. Astley, S. B., Elliott, R. M., Archer, D. B. & Southon, S. Evidence that dietary supplementation with carotenoids and carotenoid-rich foods modulates the DNA damage:repair balance in human lymphocytes. *Br. J. Nutr.* **91**, 63–72 (2004).

16. Sheng, Y., Pero, R. W., Olsson, A. R., Bryngelsson, C. & Hua, J. DNA repair enhancement by a combined supplement of carotenoids, nicotinamide, and zinc. *Cancer Detect. Prev.* **22**, 284–92 (1998).

17. Qin, J. J. *et al.* Natural products targeting the p53-MDM2 pathway and mutant p53: Recent advances and implications in cancer medicine. *Genes and Diseases* vol. 5 204–219 (2018).

18. Basten, G. P. *et al.* Sensitivity of markers of DNA stability and DNA repair activity to folate supplementation in healthy volunteers. *Br. J. Cancer* **94**, 1942–1947 (2006).

19. Connelly-Frost, A. *et al.* Selenium, folate, and colon cancer. *Nutr. Cancer* **61**, 165–178 (2009).

20. Which Fruits & Vegetables Boost DNA Repair? | NutritionFacts.org. https://nutritionfacts.org/video/fruits-vegetables-boost-dna-repair/.

21. Veggies Contain Chemicals That Boost DNA Repair And Protect Against Cancer -- ScienceDaily. https://www.sciencedaily.com/releases/2006/02/060209185153.htm.

22. Fogarty, M. C., Hughes, C. M., Burke, G., Brown, J. C. & Davison, G. W. Acute and chronic watercress supplementation attenuates exercise-induced peripheral mononuclear cell DNA damage and lipid peroxidation. *Br. J. Nutr.* **109**, 293–301 (2013).

23. Fan, S., Meng, Q., Auborn, K., Carter, T. & Rosen, E. M. BRCA1 and BRCA2 as molecular targets for phytochemicals indole-3-carbinol and genistein in breast and prostate cancer cells. *Br. J. Cancer* **94**, 407–426 (2006).

24. Hassan, F. U. *et al.* Curcumin as an alternative epigenetic modulator: Mechanism of action and potential effects. *Front. Genet.* **10**, 514 (2019).

25. Thakur, V. S., Gupta, K. & Gupta, S. Green tea polyphenols causes cell cycle arrest and apoptosis in prostate cancer cells by suppressing class I histone deacetylases. *Carcinogenesis* **33**, 377–384 (2012).

26. Balu, M., Sangeetha, P., Murali, G. & Panneerselvam, C. Modulatory role of grape seed extract on age-related oxidative DNA damage in central nervous system of rats. *Brain Res. Bull.* **68**, 469–473 (2006).

27. Balu, M., Sangeetha, P., Haripriya, D. & Panneerselvam, C. Rejuvenation of antioxidant system in central nervous system of aged rats by grape seed extract. *Neurosci. Lett.* **383**, 295–300 (2005).

28. Grabowska, W., Sikora, E. & Bielak-Zmijewska, A. Sirtuins, a promising target in slowing down the ageing process. *Biogerontology* vol. 18 447–476 (2017).

29. Rajman, L., Chwalek, K. & Sinclair, D. A. Therapeutic Potential of NAD-Boosting Molecules: The In Vivo Evidence. *Cell Metabolism* vol. 27 529–547 (2018).

30. Nicolson, G. L. Mitochondrial dysfunction and chronic disease: Treatment with natural supplements. *Integr. Med.* **13**, 35–43 (2014).

31. Chiang, S. C. *et al.* Mitochondrial protein-linked DNA breaks perturb mitochondrial gene transcription and trigger free radical–induced DNA damage. *Sci. Adv.* **3**, (2017).

32. Duberley, K. E. C. *et al.* Human neuronal coenzyme Q10 deficiency results in global loss of mitochondrial respiratory chain activity, increased mitochondrial oxidative stress and reversal of ATP synthase activity: Implications for pathogenesis & treatment. *J. Inherit. Metab. Dis.* **36**, 63–73 (2013).

33. Misra, H. S., Rajpurohit, Y. S. & Khairnar, N. P. Pyrroloquinoline-quinone and its versatile roles in biological processes. *Journal of Biosciences* vol. 37 313–325 (2012).

34. Parikh, S. *et al.* Diagnosis and management of mitochondrial disease: A consensus statement from the Mitochondrial Medicine Society. *Genetics in Medicine* vol. 17 689–701 (2015).

35. Tian, G. *et al.* Ubiquinol-10 supplementation activates mitochondria functions to decelerate senescence in senescence-accelerated mice. *Antioxidants Redox Signal.* **20**, 2606–2620 (2014).

36. Marcoff, L. & Thompson, P. D. The Role of Coenzyme Q10 in Statin-Associated Myopathy. A Systematic Review. *Journal of the American College of Cardiology* vol. 49 2231–2237 (2007).

37. Sims, C. A. *et al.* Nicotinamide mononucleotide preserves mitochondrial function and increases survival in hemorrhagic shock. *JCI insight* **3**, (2018).

38. Eckert, G. P. *et al.* Plant derived omega-3-fatty acids protect mitochondrial function in the brain. *Pharmacol. Res.* **61**, 234–241 (2010).

39. de Oliveira, M. R., Jardim, F. R., Setzer, W. N., Nabavi, S. M. & Nabavi, S. F. Curcumin, mitochondrial biogenesis, and mitophagy: Exploring recent data and indicating future needs. *Biotechnology Advances* vol. 34 813–826 (2016).

40.   Oliviero, F. *et al.* FRI0036 Epigallocatechin gallate modulates SIRT1 expression in CPP crystal-induced inflammation. *Ann. Rheum. Dis.* **71**, 321.3-322 (2013).

41.   Singh, B., Schoeb, T. R., Bajpai, P., Slominski, A. & Singh, K. K. Reversing wrinkled skin and hair loss in mice by restoring mitochondrial function. *Cell Death Dis.* **9**, 1–14 (2018).

42.   Kervezee, L., Cuesta, M., Cermakian, N. & Boivin, D. B. Simulated night shift work induces circadian misalignment of the human peripheral blood mononuclear cell transcriptome. *Proc. Natl. Acad. Sci. U. S. A.* **115**, 5540–5545 (2018).

43.   Wolff, G. L., Kodell, R. L., Moore, S. R. & Cooney, C. A. Maternal epigenetics and methyl supplements affect agouti gene expression in A vy /a mice . *FASEB J.* **12**, 949–957 (1998).

44.   Wilke, B. C. *et al.* Selenium, glutathione peroxidase (GSH-Px) and lipid peroxidation products before and after selenium supplementation. *Clin. Chim. Acta* **207**, 137–142 (1992).

45.   Piper, J. T. *et al.* Mechanisms of anticarcinogenic properties of curcumin: The effect of curcumin on glutathione linked detoxification enzymes in rat liver. *Int. J. Biochem. Cell Biol.* **30**, 445–456 (1998).

46.   Hanif, R., Qiao, L., Shiff, S. J. & Rigas, B. Curcumin, a natural plant phenolic food additive, inhibits cell proliferation and induces cell cycle changes in colon adenocarcinoma cell lines by a prostaglandin-independent pathway. *J. Lab. Clin. Med.* **130**, 576–584 (1997).

47.   Ornish, D. *et al.* Intensive lifestyle changes may affect the progression of prostate cancer. *J. Urol.* **174**, 1065–1070 (2005).

48.   Hever, J. & Cronise, R. J. Plant-based nutrition for healthcare professionals: implementing diet as a primary modality in the prevention and treatment of chronic disease. *J. Geriatr. Cardiol.* **14**, 355–368 (2017).

49.   Wang, J. *et al.* Epigenetic modulation of inflammation and synaptic plasticity promotes resilience against stress in mice. *Nat. Commun.* **9**, (2018).

50.   Kim, M. *et al.* Comparison of Blueberry (Vaccinium spp.) and Vitamin C via Antioxidative and Epigenetic Effects in Human. *J. Cancer Prev.* **22**, 174–181 (2017).

51.   Lindholm, M. *The search for human skeletal muscle memory exercise effects on the transcriptome and epigenome.* (Karolinska Institutet, 2015).

52.   Krautkramer, K. A. *et al.* Diet-Microbiota Interactions Mediate Global Epigenetic Programming in Multiple Host Tissues. *Mol. Cell* **64**, 982–992 (2016).

53. Levy, M., Thaiss, C. A. & Elinav, E. Metabolites: messengers between the microbiota and the immune system. *Genes Dev.* **30**, 1589–97 (2016).

54. Levy, M., Blacher, E. & Elinav, E. Microbiome, metabolites and host immunity. *Current Opinion in Microbiology* vol. 35 8–15 (2017).

55. Szentirmai, É., Millican, N. S., Massie, A. R. & Kapás, L. Butyrate, a metabolite of intestinal bacteria, enhances sleep. *Sci. Rep.* **9**, 1–9 (2019).

56. Biology of Aging Lotta Granholm Center on Aging MUSC. - ppt download. https://slideplayer.com/slide/4651714/.

57. Østhus, I. B. Ø. *et al.* Telomere length and long-term endurance exercise: does exercise training affect biological age? A pilot study. *PLoS One* **7**, e52769 (2012).

58. Tawani, A. & Kumar, A. Structural Insight into the interaction of Flavonoids with Human Telomeric Sequence. *Sci. Rep.* **5**, 1–13 (2015).

59. Guasch-Ferré, M. *et al.* Olive oil intake and risk of cardiovascular disease and mortality in the PREDIMED Study. *BMC Med.* **12**, 78 (2014).

60. Covas, M. I. Olive oil and the cardiovascular system. *Pharmacological Research* vol. 55 175–186 (2007).

61. Tucker, L. A. Consumption of nuts and seeds and telomere length in 5,582 men and women of the National Health and Nutrition Examination Survey (NHANES). *J. Nutr. Heal. Aging* **21**, 233–240 (2017).

62. Guasch-Ferré, M. *et al.* Frequency of nut consumption and mortality risk in the PREDIMED nutrition intervention trial. *BMC Med.* **11**, 164 (2013).

63. Callaway, E. Telomerase reverses ageing process. *Nature* (2010) doi:10.1038/news.2010.635.

64. Bickforb, P. C. *et al.* Nutraceuticals synergistically promote proliferation of human stem cells. *Stem Cells Dev.* **15**, 118–123 (2006).

65. Wolf, I. *et al.* Klotho: A tumor suppressor and a modulator of the IGF-1 and FGF pathways in human breast cancer. *Oncogene* **27**, 7094–7105 (2008).

66. Delcroix, V. *et al.* The role of the anti-aging protein klotho in IGF-1 signaling and reticular calcium leak: Impact on the chemosensitivity of dedifferentiated liposarcomas. *Cancers (Basel).* **10**, (2018).

67. MacAluso, F. & Myburgh, K. H. Current evidence that exercise can increase the number of adult stem cells. *J. Muscle Res. Cell Motil.* **33**, 187–198 (2012).

68. Blackmore, D. G., Golmohammadi, M. G., Large, B., Waters, M. J. & Rietze, R. L. Exercise increases neural stem cell number in a growth hormone-

dependent manner, augmenting the regenerative response in aged mice. *Stem Cells* **27**, 2044–2052 (2009).

69. Ceccarelli, G., Benedetti, L., Arcari, M. L., Carubbi, C. & Galli, D. Muscle stem cell and physical activity: What point is the debate at? *Open Medicine (Poland)* vol. 12 144–156 (2017).

70. Effect of exercise on stem cells. https://www.brighton.ac.uk/crmd/what-we-do/research-projects/effect-of-exercise-on-stem-cells.aspx.

71. Exercise boosts health by influencing stem cells to become bone, not fat, researchers find. https://phys.org/news/2011-09-boosts-health-stem-cells-bone.html.

72. Buettner, D. & Skemp, S. Blue Zones: Lessons From the World's Longest Lived. *American Journal of Lifestyle Medicine* vol. 10 318–321 (2016).

73. Babu, P. V. A., Sabitha, K. E. & Shyamaladevi, C. S. Effect of green tea extract on advanced glycation and cross-linking of tail tendon collagen in streptozotocin induced diabetic rats. *Food Chem. Toxicol.* **46**, 280–285 (2008).

74. Jariyapamornkoon, N., Yibchok-anun, S. & Adisakwattana, S. Inhibition of advanced glycation end products by red grape skin extract and its antioxidant activity. *BMC Complement. Altern. Med.* **13**, 171 (2013).

75. Holzenberger, M. *et al.* IGF-1 receptor regulates lifespan and resistance to oxidative stress in mice. *Nature* **421**, 182–187 (2003).

76. Junnila, R.K., List, E.O., Berryman, D.E., Murrey, J.W. & Kopchick, J.J. The GH/IGF-1 axis in ageing & longevity. *Nature Reviews Endocrin.* vol. 9 366–376 (2013).

77. (PDF) Prebiotic foods in association with plasma levels of IGF-1 and IGFBP-3 in breast cancer patients. https://www.researchgate.net/publication/278405307_Prebiotic_foods_in_association_with_plasma_levels_of_IGF-1_and_IGFBP-3_in_breast_cancer_patients.

78. Viollet, B. *et al.* Cellular and molecular mechanisms of metformin: An overview. *Clinical Science* vol. 122 253–270 (2012).

79. Barzilai, N., Crandall, J. P., Kritchevsky, S. B. & Espeland, M. A. Metformin as a Tool to Target Aging. *Cell Metabolism* vol. 23 1060–1065 (2016).

80. Konopka, A. R. *et al.* Metformin inhibits mitochondrial adaptations to aerobic exercise training in older adults. *Aging Cell* **18**, (2019).

81. Aroda, V. R. *et al.* Long-term metformin use and vitamin B12 deficiency in the diabetes prevention program outcomes study. *J. Clin. Endocrinol. Metab.* **101**, 1754–1761 (2016).

82. Yin, J., Xing, H. & Ye, J. Efficacy of berberine in patients with type 2 diabetes mellitus. *Metabolism.* **57**, 712–717 (2008).

83.   Ma, X. *et al.* The Pathogenesis of Diabetes Mellitus by Oxidative Stress and Inflammation: Its Inhibition by Berberine. *Front. Pharmacol.* **9**, 782 (2018).

84.   Sabatini, D. M. Twenty-five years of mTOR: Uncovering the link from nutrients to growth. *Proceedings of the National Academy of Sciences of the United States of America* vol. 114 11818–11825 (2017).

85.   3 Relatively Unknown Protein-Related Problems (And How to Fix Them) - UC Davis. https://ucdintegrativemedicine.com/2016/10/3-relatively-unknown-protein-related-problems-fix/#gs.5x662i.

86.   Uh, D. *et al.* Dehydroepiandrosterone sulfate level varies nonlinearly with symptom severity in major depressive disorder. *Clin. Psychopharmacol. Neurosci.* **15**, 163–169 (2017).

87.   Arlt, W. Dehydroepiandrosterone and ageing. *Best Pract. Res. Clin. Endocrinol. Metab.* **18**, 363–380 (2004).

88.   De Groot, S., Pijl, H., Van Der Hoeven, J. J. M. & Kroep, J. R. Effects of short-term fasting on cancer treatment. *Journal of Experimental and Clinical Cancer Research* vol. 38 (2019).

89.   Richter, E. A. & Ruderman, N. B. AMPK and the biochemistry of exercise: Implications for human health and disease. *Biochemical Journal* vol. 418 261–275 (2009).

90.   Hardie, D. G. Energy sensing by the AMP-activated protein kinase and its effects on muscle metabolism. in *Proceedings of the Nutrition Society* vol. 70 92–99 (2011).

91.   Chang, H. C. & Guarente, L. SIRT1 and other sirtuins in metabolism. *Trends in Endocrinology and Metabolism* vol. 25 138–145 (2014).

92.   Stead, E. R. *et al.* Agephagy – Adapting Autophagy for Health During Aging. *Frontiers in Cell and Developmental Biology* vol. 7 (2019).

93.   Logan, J. Dyslexic entrepreneurs: the incidence; their coping strategies and their business skills. *Dyslexia* **15**, 328–346 (2009).

94.   Van Cauter, E. & Plat, L. Physiology of growth hormone secretion during sleep. in *Journal of Pediatrics* vol. 128 (J Pediatr, 1996).

95.   Cauter, E. Van *et al.* A quantitative estimation of growth hormone secretion in normal man: Reproducibility and relation to sleep and time of day. *J. Clin. Endocrinol. Metab.* **74**, 1441–1450 (1992).

96.   Wahl, P., Zinner, C., Achtzehn, S., Bloch, W. & Mester, J. Effect of high- and low-intensity exercise and metabolic acidosis on levels of GH, IGF-I, IGFBP-3 and cortisol. *Growth Horm. IGF Res.* **20**, 380–385 (2010).

97.   Leppäluoto, J. *et al.* Endocrine effects of repeated sauna bathing. *Acta Physiol. Scand.* **128**, 467–70 (1986).

98.   Rando, T. A. & Chang, H. Y. Aging, rejuvenation, and epigenetic reprogramming: Resetting the aging clock. *Cell* vol. 148 46–57 (2012).

99.   Kaufman, R. J. *et al.* The unfolded protein response in nutrient sensing and differentiation. *Nature Reviews Molecular Cell Biology* vol. 3 411–421 (2002).

100.   Stefani, M. & Rigacci, S. Protein folding and aggregation into amyloid: The interference by natural phenolic compounds. *International Journal of Molecular Sciences* vol. 14 12411–12457 (2013).

101.   Moura, C. S., Lollo, P. C. B., Morato, P. N. & Amaya-Farfan, J. Dietary nutrients and bioactive substances modulate heat shock protein (HSP) expression: A review. *Nutrients* vol. 10 (2018).

102.   Giunta, B. *et al.* Fish oil enhances anti-amyloidogenic properties of green tea EGCG in Tg2576 mice. *Neurosci. Lett.* **471**, 134–138 (2010).

103.   Yang, F. *et al.* Curcumin inhibits formation of amyloid β oligomers and fibrils, binds plaques, and reduces amyloid in vivo. *J. Biol. Chem.* **280**, 5892–5901 (2005).

104.   Maiti, P., Manna, J., Veleri, S., Frautschy, S. Molecular Chaperone Dysfunction in Neurodegenerative Diseases and Effects of Curcumin. *Biomed Res. Int.* (2014).

105.   Calapai, G. *et al.* A randomized, double-blinded, clinical trial on effects of a Vitis vinifera extract on cognitive function in healthy older adults. *Front. Pharmacol.* **8**, (2017).

106.   Allam, F. *et al.* Grape Powder Supplementation Prevents Oxidative Stress–Induced Anxiety-Like Behavior, Memory Impairment, and High Blood Pressure in Rats. *J. Nutr.* **143**, 835–842 (2013).

107.   Sano, A. *et al.* Beneficial effects of grape seed extract on malondialdehyde-modified LDL. *J. Nutr. Sci. Vitaminol. (Tokyo).* **53**, 174–182 (2007).

108.   Bagchi, D. *et al.* Molecular mechanisms of cardioprotection by a novel grape seed proanthocyanidin extract. in *Mutation Research - Fundamental and Molecular Mechanisms of Mutagenesis* vols 523–524 87–97 (Elsevier, 2003).

109.   Lupoli, R. *et al.* Impact of Grape Products on Lipid Profile: A Meta-Analysis of Randomized Controlled Studies. *J. Clin. Med.* **9**, 313 (2020).

110.   Derry, M., Raina, K., Agarwal, R. & Agarwal, C. Differential effects of grape seed extract against human colorectal cancer cell lines: The intricate role of death receptors and mitochondria. *Cancer Lett.* **334**, 69–78 (2013).

111. Liu, Y. *et al.* Gallic acid is the major component of grape seed extract that inhibits amyloid fibril formation. *Bioorganic Med. Chem. Lett.* **23**, 6336–6340 (2013).

112. Jayamani, J. & Shanmugam, G. Gallic acid, one of the components in many plant tissues, is a potential inhibitor for insulin amyloid fibril formation. *Eur. J. Med. Chem.* **85**, 352–8 (2014).

113. Kim, H. *et al.* Effects of naturally occurring compounds on fibril formation and oxidative stress of β-amyloid. *J. Agric. Food Chem.* **53**, 8537–8541 (2005).

114. Berr, C. *et al.* Olive oil and cognition: Results from the three-city study. *Dement. Geriatr. Cogn. Disord.* **28**, 357–364 (2009).

115. Mazza, E. *et al.* Effect of the replacement of dietary vegetable oils with a low dose of extravirgin olive oil in the Mediterranean Diet on cognitive functions in the elderly. *J. Transl. Med.* **16**, 10 (2018).

116. Barbaro, B. *et al.* Effects of the olive-derived polyphenol oleuropein on human health. *International Journal of Molecular Sciences* vol. 15 18508–18524 (2014).

117. Chaudhuri, T. K. & Paul, S. Protein-misfolding diseases and chaperone-based therapeutic approaches. *FEBS Journal* vol. 273 1331–1349 (2006).

118. Agorogiannis, E. I., Agorogiannis, G. I., Papadimitriou, A. & Hadjigeorgiou, G. M. Protein misfolding in neurodegenerative diseases. *Neuropathology and Applied Neurobiology* vol. 30 215–224 (2004).

119. Morton, J. P., Kayani, A. C., McArdle, A. & Drust, B. The Exercise-Induced stress response of skeletal muscle, with specific emphasis on humans. *Sports Medicine* vol. 39 643–662 (2009).

120. Laukkanen, T., Khan, H., Zaccardi, F. & Laukkanen, J. A. Association between sauna bathing and fatal cardiovascular and all-cause mortality events. *JAMA Intern. Med.* **175**, 542–548 (2015).

121. Freitas-Rodríguez, S., Folgueras, A. R. & López-Otín, C. The role of matrix metalloproteinases in aging: Tissue remodeling and beyond. *Biochimica et Biophysica Acta - Molecular Cell Research* vol. 1864 2015–2025 (2017).

122. Itoh, Y. & Nagase, H. Matrix metalloproteinases in cancer. *Essays Biochem.* **38**, 21–36 (2002).

123. Quintero-Fabián, S. *et al.* Role of Matrix Metalloproteinases in Angiogenesis and Cancer. *Frontiers in Oncology* vol. 9 1370 (2019).

124. Piperigkou, Z., Manou, D., Karamanou, K. & Theocharis, A. D. Strategies to target matrix metalloproteinases as therapeutic approach in cancer. in *Methods in Molecular Biology* vol. 1731 325–348 (Humana Press Inc., 2018).

125. Gupta, P. Natural Products as Inhibitors of Matrix Metalloproteinases. Nat Prod Chem Res 4: e114. References 1. Nagese H, Woessner JF (1999) Matrix Metalloproteinases. *J. Biol. Chem.* **4**, 21491–21494 (2016).

126. Mukherjee, P. K., Maity, N., Nema, N. K. & Sarkar, B. K. Natural matrix metalloproteinase inhibitors: Leads from herbal resources. in *Studies in Natural Products Chemistry* vol. 39 91–113 (Elsevier B.V., 2013).

127. Zhang, C. & Kim, S. K. Matrix metalloproteinase inhibitors (MMPIs) from marine natural products: The current situation and future prospects. *Marine Drugs* vol. 7 71–84 (2009).

128. Thaiss, C. A., Zmora, N., Levy, M. & Elinav, E. The microbiome and innate immunity. *Nature* vol. 535 65–74 (2016).

129. Foster, J. A. & McVey Neufeld, K. A. Gut-brain axis: How the microbiome influences anxiety and depression. *Trends in Neurosciences* vol. 36 305–312 (2013).

130. Microbes Help Produce Serotonin in Gut | www.caltech.edu. https://www.caltech.edu/about/news/microbes-help-produce-serotonin-gut-46495.

131. The Brain-Gut Connection | Johns Hopkins Medicine. https://www.hopkinsmedicine.org/health/wellness-and-prevention/the-brain-gut-connection.

132. Deng, F., Li, Y. & Zhao, J. The gut microbiome of healthy long-living people. *Aging* vol. 11 289–290 (2019).

133. Leaky gut: What is it, and what does it mean for you? - Harvard Health Blog - Harvard Health Publishing. https://www.health.harvard.edu/blog/leaky-gut-what-is-it-and-what-does-it-mean-for-you-2017092212451.

134. Do, M. H., Lee, E., Oh, M. J., Kim, Y. & Park, H. Y. High-glucose or-fructose diet cause changes of the gut microbiota and metabolic disorders in mice without body weight change. *Nutrients* **10**, (2018).

135. Allen, R. J. & Waclaw, B. Bacterial growth: a statistical physicist's guide. *Rep. Prog. Phys.* **82**, 016601 (2019).

136. Shane, A. L. Missing Microbes: How the Overuse of Antibiotics Is Fueling Our Modern Plagues. *Emerg. Infect. Dis.* **20**, 1961–1961 (2014).

137. 'Missing Microbes': How Antibiotics Can Do Harm - The New York Times. https://www.nytimes.com/2014/04/29/health/missing-microbes-how-antibiotics-can-do-harm.html.

138. Lach, G., Schellekens, H., Dinan, T. G. & Cryan, J. F. Anxiety, Depression, and the Microbiome: A Role for Gut Peptides. *Neurotherapeutics* vol. 15 36–59 (2018).

139. Peirce, J. M. & Alviña, K. The role of inflammation and the gut microbiome

in depression and anxiety. *Journal of Neuroscience Research* vol. 97 1223–1241 (2019).

140.	Rieder, R., Wisniewski, P. J., Alderman, B. L. & Campbell, S. C. Microbes and mental health: A review. *Brain, Behavior, and Immunity* vol. 66 9–17 (2017).

141.	Bruce-Keller, A. J., Salbaum, J. M. & Berthoud, H. R. Harnessing Gut Microbes for Mental Health: Getting From Here to There. *Biological Psychiatry* vol. 83 214–223 (2018).

142.	Pinto-Sanchez, M. I. *et al.* Probiotic Bifidobacterium longum NCC3001 Reduces Depression Scores and Alters Brain Activity: A Pilot Study in Patients With Irritable Bowel Syndrome. *Gastroenterology* **153**, 448-459.e8 (2017).

143.	Akbari, E. *et al.* Effect of probiotic supplementation on cognitive function and metabolic status in Alzheimer's disease: A randomized, double-blind and controlled trial. *Front. Aging Neurosci.* **8**, (2016).

144.	Parkinson's Disease Linked to Microbiome | www.caltech.edu. https://www.caltech.edu/about/news/parkinsons-disease-linked-microbiome-53109.

145.	Rao, A. V. *et al.* A randomized, double-blind, placebo-controlled pilot study of a probiotic in emotional symptoms of chronic fatigue syndrome. *Gut Pathog.* **1**, 6 (2009).

146.	Kuenzig, M. E., Bishay, K., Leigh, R., Kaplan, G. G. & Benchimol, E. I. Co-occurrence of Asthma and the Inflammatory Bowel Diseases: A Systematic Review and Meta-analysis. *Clinical and Translational Gastroenterology* vol. 9 (2018).

147.	Miller, L. E., Lehtoranta, L. & Lehtinen, M. J. Short-term probiotic supplementation enhances cellular immune function in healthy elderly: systematic review and meta-analysis of controlled studies. *Nutrition Research* vol. 64 1–8 (2019).

148.	Jäger, R., Purpura, M., Farmer, S., Cash, H. A. & Keller, D. Probiotic Bacillus coagulans GBI-30, 6086 improves protein absorption and utilization. *Probiotics Antimicrob. Proteins* **10**, 611–615 (2018).

149.	Wang, L. *et al.* The effects of probiotics on total cholesterol. *Med. (United States)* **97**, (2018).

150.	Khanna, S. *et al.* Changes in microbial ecology after fecal microbiota transplantation for recurrent C. difficile infection affected by underlying inflammatory bowel disease. *Microbiome* **5**, 1–8 (2017).

151.	Weingarden, A. *et al.* Dynamic changes in short- and long-term bacterial composition following fecal microbiota transplantation for recurrent Clostridium difficile infection. *Microbiome* **3**, 10 (2015).

152.   Górska, A., Przystupski, D., Niemczura, M. J. & Kulbacka, J. Probiotic Bacteria: A Promising Tool in Cancer Prevention and Therapy. *Current Microbiology* vol. 76 939–949 (2019).

153.   Collins, F. L., Rios-Arce, N. D., Schepper, J. D., Parameswaran, N. & McCabe, L. R. The Potential of Probiotics as a Therapy for Osteoporosis. *Microbiol. Spectr.* **5**, (2017).

154.   Lambert, M. N. T. *et al.* Combined bioavailable isoflavones and probiotics improve bone status and estrogen metabolism in postmenopausal osteopenic women: A randomized controlled trial. *Am. J. Clin. Nutr.* **106**, 909–920 (2017).

155.   Wang, J. *et al.* Modulation of gut microbiota during probiotic-mediated attenuation of metabolic syndrome in high fat diet-fed mice. *ISME J.* **9**, 1–15 (2014).

156.   Sanchez, M. *et al.* Effect of Lactobacillus rhamnosus CGMCC1.3724 supplementation on weight loss and maintenance in obese men and women. *Br. J. Nutr.* **111**, 1507–1519 (2014).

157.   Zhang, Q., Wu, Y. & Fei, X. Effect of probiotics on body weight and body-mass index: a systematic review and meta-analysis of randomized, controlled trials. *Int. J. Food Sci. Nutr.* **67**, 571–580 (2016).

158.   Exercise boosts well-being by improving gut health. https://www.medicalnewstoday.com/articles/324465#1.

159.   Montecino-Rodriguez, E., Berent-Maoz, B. & Dorshkind, K. Causes, consequences, and reversal of immune system aging. *Journal of Clinical Investigation* vol. 123 958–965 (2013).

160.   2009 Swine-Flu Death Toll 10 Times Higher Than Thought | Live Science. https://www.livescience.com/41539-2009-swine-flu-death-toll-higher.html.

161.   Micronutrients have major impact on health - Harvard Health. https://www.health.harvard.edu/staying-healthy/micronutrients-have-major-impact-on-health.

162.   Bou Ghanem, E. N. *et al.* The α-Tocopherol Form of Vitamin E Reverses Age-Associated Susceptibility to *Streptococcus pneumoniae* Lung Infection by Modulating Pulmonary Neutrophil Recruitment. *J. Immunol.* **194**, 1090–1099 (2015).

163.   Muriach, M. *et al.* Lutein prevents the effect of high glucose levels on immune system cells in vivo and in vitro. *J. Physiol. Biochem.* **64**, 149–157 (2008).

164.   Immunomodulation and Anti-Inflammatory Effects of Garlic Compounds. https://www.ncbi.nlm.nih.gov/pmc/articles/PMC4417560/.

165.   Lee, G. Y. & Han, S. N. Role of vitamin E in immunity. *Nutrients* vol. 10 (2018).

166.   Guggenheim, A. G., Wright, K. M. & Zwickey, H. L. Immune modulation from five major mushrooms: Application to integrative oncology. *Integrative Medicine (Boulder)* vol. 13 32–44 (2014).

167.   Fucoidan | Memorial Sloan Kettering Cancer Center. https://www.mskcc.org/cancer-care/integrative-medicine/herbs/fucoidan#msk_professional.

168.   Luthuli, S. *et al.* Therapeutic effects of fucoidan: A review on recent studies. *Marine Drugs* vol. 17 (2019).

169.   Vetvicka, V., Vannucci, L., Sima, P. & Richter, J. Beta glucan: Supplement or drug? From laboratory to clinical trials. *Molecules* vol. 24 (2019).

170.   Geller, A., Shrestha, R. & Yan, J. Yeast-derived β-glucan in cancer: Novel uses of a traditional therapeutic. *Int Journal of Molecular Sciences* vol. 20 (2019).

171.   Hong, F. *et al.* Mechanism by Which Orally Administered β-1,3-Glucans Enhance the Tumoricidal Activity of Antitumor Monoclonal Antibodies in Murine Tumor Models. *J. Immunol.* **173**, 797–806 (2004).

172.   Postirradiation Glucan Administration Enhances the Radioprotective Effects of WR-2721 - PubMed. https://pubmed.ncbi.nlm.nih.gov/2536480/.

173.   Survival Enhancement and Hemopoietic Regeneration Following Radiation Exposure: Therapeutic Approach Using Glucan and Granulocyte Colony-Stimulating Factor - PubMed. https://pubmed.ncbi.nlm.nih.gov/1697806/.

174.   β-Glucan Functions as an Adjuvant for Monoclonal Antibody Immunotherapy by Recruiting Tumoricidal Granulocytes as Killer Cells | Cancer Research. https://cancerres.aacrjournals.org/content/63/24/9023.

175.   Hardy, H., Harris, J., Lyon, E., Beal, J. & Foey, A. D. Probiotics, prebiotics and immunomodulation of gut mucosal defences: Homeostasis and immunopathology. *Nutrients* vol. 5 1869–1912 (2013).

176.   Miller, M. B. & Bassler, B. L. Quorum Sensing in Bacteria. *Annu. Rev. Microbiol.* **55**, 165–199 (2001).

177.   Reading, N. C. & Sperandio, V. Quorum sensing: The many languages of bacteria. *FEMS Microbiology Letters* vol. 254 1–11 (2006).

178.   Brackman, G., Cos, P., Maes, L., Nelis, H. J. & Coenye, T. Quorum sensing inhibitors increase the susceptibility of bacterial biofilms to antibiotics in vitro and in vivo. *Antimicrob. Agents Chemother.* **55**, 2655–2661 (2011).

179.   Castillo-Quan, J. I., Kinghorn, K. J. & Bjedov, I. Genetics and Pharmacology of Longevity: Road to Therapeutics for Healthy Aging. *Adv. Genet.* **90**, 1–101 (2015).

180. Brower, V. Mind-body research moves towards the mainstream. Mounting evidence for the role of the mind in disease and healing is leading to a greater acceptance of mind-body medicine. *EMBO Rep.* **7**, 358–361 (2006).

181. Longevity Increased by Positive Self-Perceptions of Aging - PubMed. https://pubmed.ncbi.nlm.nih.gov/12150226/.

182. Wikgren, M. *et al.* Short telomeres in depression and the general population are associated with a hypocortisolemic state. *Biol. Psychiatry* **71**, 294–300 (2012).

183. Depression and chronic stress accelerates aging -- ScienceDaily. https://www.sciencedaily.com/releases/2011/11/111109093729.htm.

184. Barton, J. & Pretty, J. What is the best dose of nature and green exercise for improving mental health- A multi-study analysis. *Environ. Sci. Technol.* **44**, 3947–3955 (2010).

185. Davis, D. M. & Hayes, J. A. What Are the Benefits of Mindfulness? A Practice Review of Psychotherapy-Related Research. *Psychotherapy* vol. 48 198–208 (2011).

186. Epel, E., Daubenmier, J., Moskowitz, J.T., Folkman, S. & Blackburn, E. Can meditation slow rate of cellular aging? Cognitive stress, mindfulness, and telomeres. In *Annals of the New York Academy of Sciences* vol. 1172 34-53 (Blackwell Publishing Inc., 2009).

187. Michael Mosley: 'Forget walking 10,000 steps a day' - BBC News. https://www.bbc.co.uk/news/health-42864061.

188. Gleeson, M. Immune system adaptation in elite athletes. *Current Opinion in Clinical Nutrition and Metabolic Care* vol. 9 659–665 (2006).

189. Malm, C. Susceptibility to infections in elite athletes: The S-curve. *Scandinavian Journal of Medicine and Science in Sports* vol. 16 4–6 (2006).

190. Tabara, Y. *et al.* Association of postural instability with asymptomatic cerebrovascular damage and cognitive decline: The japan shimanami health promoting program study. *Stroke* **46**, 16–22 (2015).

191. How To Improve Your Balance | Memory Foundation. https://www.memory.foundation/2015/05/18/how-to-improve-your-balance/.

192. Barbara Springer, C. A., Raul Marin, C., Cyhan, T. & Springer, B. A. *Normative Values for the Unipedal Stance Test with Eyes Open and Closed. Journal of Geriatric Physical Therapy* vol. 30.

193. Kubota, Y., Alonso, A., Shah, A. M., Chen, L. Y. & Folsom, A. R. Television watching as sedentary behavior and atrial fibrillation: The atherosclerosis risk in communities study. *J. Phys. Act. Heal.* **15**, 895–899 (2018).

194. Every hour of TV watching shortens life by 22 minutes - Telegraph. https://www.telegraph.co.uk/news/health/news/8702101/Every-hour-of-TV-watching-shortens-life-by-22-minutes.html.

195. Feinman, R. D. & Fine, E. J. 'A calorie is a calorie' violates the second law of thermodynamics. *Nutrition Journal* vol. 3 9 (2004).

196. Feinman, R. D. & Fine, E. J. Thermodynamics and Metabolic Advantage of Weight Loss Diets. *Metab. Syndr. Relat. Disord.* **1**, 209–219 (2003).

197. Jakubowicz, D., Barnea, M., Wainstein, J. & Froy, O. High Caloric intake at breakfast vs. dinner differentially influences weight loss of overweight and obese women. *Obesity* **21**, 2504–2512 (2013).

198. Raynor, H. A., Li, F. & Cardoso, C. Daily pattern of energy distribution and weight loss. *Physiol. Behav.* **192**, 167–172 (2018).

199. Kahleova, H., Lloren, J. I., Mashchak, A., Hill, M. & Fraser, G. E. Meal Frequency and Timing Are Associated with Changes in Body Mass Index in Adventist Health Study 2. *J. Nutr.* **147**, jn244749 (2017).

200. Colman, R. J. *et al.* Caloric restriction delays disease onset and mortality in rhesus monkeys. *Science (80-.).* **325**, 201–204 (2009).

201. Calorie Restriction and Fasting Diets: What Do We Know? | National Institute on Aging. https://www.nia.nih.gov/health/calorie-restriction-and-fasting-diets-what-do-we-know.

202. Part 6: Diet - NHS Digital. https://digital.nhs.uk/data-and-information/publications/statistical/statistics-on-obesity-physical-activity-and-diet/statistics-on-obesity-physical-activity-and-diet-england-2019/part-6-diet.

203. American Cancer Society Guidelines on Nutrition and Physical Activity for Cancer Prevention Summary of the ACS Guidelines on Nutrition and Physical Activity. doi:10.3322/caac.20140/full.

204. Fruit and vegetable intake and the risk of cardiovascular disease, total cancer and all-cause mortality | Int J. of Epidemiology | Oxford Academic. https://academic.oup.com/ije/article/46/3/1029/3039477.

205. Forget five a day, eat 10 portions of fruit and veg to cut risk of early death | Society | The Guardian https://www.theguardian.com/society/2017/feb/23/five-day-10-portions-fruit-veg-cut-early-death.

206. Anti-inflammatory Properties of Curcumin, a Major Constituent of Curcuma Longa: A Review of Preclinical and Clinical Research - PubMed. https://pubmed.ncbi.nlm.nih.gov/19594223/.

207.   Biswas, S. K., McClure, D., Jimenez, L. A., Megson, I. L. & Rahman, I. Curcumin induces glutathione biosynthesis & inhibits NF-κB activation and interleukin-8 release in alveolar epithelial cells: Mechanism of free radical scavenging activity. *Antioxidants and Redox Signaling* vol. 7 32–41 (2005).

208.   Wongcharoen, W. & Phrommintikul, A. The protective role of curcumin in cardiovascular diseases. *Int Journal of Cardiology* vol. 133 145–151 (2009).

209.   Usharani, P., Mateen, A. A., Naidu, M. U. R., Raju, Y. S. N. & Chandra, N. Effect of NCB-02, Atorvastatin and Placebo on Endothelial Function, Oxidative Stress and Inflammatory Markers in Patients with Type 2 Diabetes Mellitus: A Randomized, Parallel-Group, Placebo-Controlled, 8-Week Study. *Drugs R D* **9**, 243–250 (2008).

210.   Sanmukhani, J. *et al.* Efficacy and safety of curcumin in major depressive disorder: A randomized controlled trial. *Phyther. Res.* **28**, 579–585 (2014).

211.   Chandran, B. & Goel, A. A randomized, pilot study to assess the efficacy and safety of curcumin in patients with active rheumatoid arthritis. *Phyther. Res.* **26**, 1719–1725 (2012).

212.   Sikora, E., Bielak-Zmijewska, A., Mosieniak, G. & Piwocka, K. The Promise of Slow Down Ageing May Come from Curcumin. *Curr. Pharm. Des.* **16**, 884–892 (2010).

213.   Shoba, G. *et al.* Influence of piperine on the pharmacokinetics of curcumin in animals and human volunteers. *Planta Med.* **64**, 353–356 (1998).

214.   Ogunleye, A. A., Xue, F. & Michels, K. B. Green tea consumption and breast cancer risk or recurrence: A meta-analysis. *Breast Cancer Res. Treat.* **119**, 477–484 (2010).

215.   Green Tea Consumption and Prostate Cancer Risk in Japanese Men: A Prospective Study | American Journal of Epidemiology | Oxford Academic. https://academic.oup.com/aje/article/167/1/71/185454.

216.   Weinreb, O., Mandel, S., Amit, T. & Youdim, M. B. H. Neurological mechanisms of green tea polyphenols in Alzheimer's and Parkinson's diseases. *Journal of Nutritional Biochemistry* vol. 15 506–516 (2004).

217.   Mandel, S. A., Amit, T., Weinreb, O., Reznichenko, L. & Youdim, M. B. H. Simultaneous manipulation of multiple brain targets by green tea catechins: A potential neuroprotective strategy for Alzheimer and Parkinson diseases. *CNS Neuroscience and Therapeutics* vol. 14 352–365 (2008).

218.   The Neuropharmacology of L-theanine(N-ethyl-L-glutamine): A Possible Neuroprotective and Cognitive Enhancing Agent - PubMed. https://pubmed.ncbi.nlm.nih.gov/17182482/.

219. Huxley, R. *et al.* Coffee, decaffeinated coffee, and tea consumption in relation to incident type 2 diabetes mellitus: A systematic review with meta-analysis. *Archives of Internal Medicine* vol. 169 2053–2063 (2009).

220. Liu, K. *et al.* Effect of green tea on glucose control and insulin sensitivity: a meta-analysis of 17 randomized controlled trials. *Am. J. Clin. Nutr.* **98**, 340–348 (2013).

221. Yokozawa, T. & Dong, E. Influence of green tea and its three major components upon low-density lipoprotein oxidation. *Exp. Toxicol. Pathol.* **49**, 329–335 (1997).

222. Kuriyama, S. *et al.* Green tea consumption and mortality due to cardiovascular disease, cancer, and all causes in Japan: The Ohsaki study. *J. Am. Med. Assoc.* **296**, 1255–1265 (2006).

223. Rauf, A. *et al.* Proanthocyanidins: A comprehensive review. *Biomedicine and Pharmacotherapy* vol. 116 108999 (2019).

224. Yang, J. & Xiao, Y. Y. Grape Phytochemicals and Associated Health Benefits. *Crit. Rev. Food Sci. Nutr.* **53**, 1202–1225 (2013).

225. Park, E., Edirisinghe, I., Choy, Y. Y., Waterhouse, A. & Burton-Freeman, B. Effects of grape seed extract beverage on blood pressure and metabolic indices in individuals with pre-hypertension: A randomised, double-blinded, two-arm, parallel, placebo-controlled trial. *Br. J. Nutr.* **115**, 226–238 (2016).

226. Prasad, R. & Katiyar, S. K. Grape seed proanthocyanidins inhibit migration potential of pancreatic cancer cells by promoting mesenchymal-to-epithelial transition and targeting NF-κB. *Cancer Lett.* **334**, 118–126 (2013).

227. Lutein and Zeaxanthin in the Diet and Serum and Their Relation to Age-Related Maculopathy in the Third National Health and Nutrition Examination Survey - PubMed. https://pubmed.ncbi.nlm.nih.gov/11226974/.

228. Kim, J. E. *et al.* A Lutein-Enriched Diet Prevents Cholesterol Accumulation and Decreases Oxidized LDL and Inflammatory Cytokines in the Aorta of Guinea Pigs. *J. Nutr.* **141**, 1458–1463 (2011).

229. Gammone, M. A., Riccioni, G. & D'Orazio, N. Carotenoids: Potential allies of cardiovascular health? *Food Nutr. Res.* **59**, (2015).

230. Assar, E. A., Vidalle, M. C., Chopra, M. & Hafizi, S. Lycopene acts through inhibition of IκB kinase to suppress NF-κB signaling in human prostate and breast cancer cells. *Tumor Biol.* **37**, 9375–9385 (2016).

231. Giovannucci, E. A review of epidemiologic studies of tomatoes, lycopene, and prostate cancer. in *Experimental Biology and Medicine* vol. 227 852–859 (Exp Biol Med (Maywood), 2002).

232. Han, G. M., Meza, J. L., Soliman, G. A., Islam, K. M. M. & Watanabe-Galloway, S. Higher levels of serum lycopene are associated with reduced mortality in individuals with metabolic syndrome. *Nutr. Res.* **36**, 402–407 (2016).

233. Soy Isoflavones | Linus Pauling Institute | Oregon State University. https://lpi.oregonstate.edu/mic/dietary-factors/phytochemicals/soy-isoflavones#cardiovascular-disease-prevention.

234. Banerjee, S., Li, Y., Wang, Z. & Sarkar, F. H. Multi-targeted therapy of cancer by genistein. *Cancer Letters* vol. 269 226–242 (2008).

235. Soy Isoflavones & Your Brain | Cognitive Vitality | Alzheimer's Drug Discovery Foundation. https://www.alzdiscovery.org/cognitive-vitality/ratings/soy-isoflavones.

236. Alekseyenko, T. V. *et al.* Antitumor and antimetastatic activity of fucoidan, a sulfated polysaccharide isolated from the Okhotsk sea Fucus evanescens brown alga. *Bull. Exp. Biol. Med.* **143**, 730–732 (2007).

237. Animal Data Shows Fisetin to be a Surprisingly Effective Senolytic – Fight Aging! https://www.fightaging.org/archives/2018/10/animal-data-shows-fisetin-to-be-a-surprisingly-effective-senolytic/.

238. Chuang, C. C. *et al.* Quercetin is equally or more effective than resveratrol in attenuating tumor necrosis factor-α-mediated inflammation and insulin resistance in primary human adipocytes. *Am. J. Clin. Nutr.* **92**, 1511–1521 (2010).

239. Yang, F. *et al.* Quercetin in prostate cancer: Chemotherapeutic and chemopreventive effects, mechanisms and clinical application potential (review). *Oncol. Rep.* **33**, 2659–2668 (2015).

240. Talalay, P. Chemoprotection against cancer by induction of Phase 2 enzymes. in *BioFactors* vol. 12 5–11 (IOS Press, 2000).

241. Shukla, S. & Gupta, S. Apigenin: A promising molecule for cancer prevention. *Pharmaceutical Research* vol. 27 962–978 (2010).

242. Vitamin C Elevates Red Blood Cell Glutathione in Healthy Adults - PubMed. https://pubmed.ncbi.nlm.nih.gov/8317379/.

243. Semba, R. D. *et al.* Resveratrol levels and all-cause mortality in older community-dwelling adults. *JAMA Intern. Med.* **174**, 1077–1084 (2014).

244. Salehi, B. *et al.* Resveratrol: A double-edged sword in health benefits. *Biomedicines* vol. 6 (2018).

245. Fiber-Rich Diet Linked to Longevity - Calorie Control Council. https://caloriecontrol.org/fiber-rich-diet-linked-to-longevity-2/.

246. Bollrath, J. & Powrie, F. Feed your Tregs more fiber. *Science* vol. 341 463–464 (2013).

247. Threapleton, D. E. *et al.* Dietary fibre intake and risk of cardiovascular disease: Systematic review and meta-analysis. *BMJ (Online)* vol. 347 (2013).

248. Morris, M. C. *et al.* MIND diet slows cognitive decline with aging. *Alzheimer's Dement.* **11**, 1015–1022 (2015).

249. Galleano, M., Oteiza, P. I. & Fraga, C. G. Cocoa, chocolate, and cardiovascular disease. *Journal of Cardiovascular Pharmacology* vol. 54 483–490 (2009).

250. Latif, R. Health benefits of cocoa. *Current Opinion in Clinical Nutrition and Metabolic Care* vol. 16 669–674 (2013).

251. Bondonno, N. P. *et al.* Flavonoid intake is associated with lower mortality in the Danish Diet Cancer and Health Cohort. *Nat. Commun.* **10**, (2019).

252. https://nutrishield.com/wp-content/uploads/2013/04/VivacellSummaryWeb.pdf.

253. Nasri, H., Baradaran, A., Shirzad, H. & Kopaei, M. R. New concepts in nutraceuticals as alternative for pharmaceuticals. *Int. J. Prev. Med.* **5**, 1487–1499 (2014).

254. Leslie, W. & Hankey, C. Aging, Nutritional Status and Health. *Healthcare* **3**, 648–658 (2015).

255. Lally, P., van Jaarsveld, C. H. M., Potts, H. W. W. & Wardle, J. How are habits formed: Modelling habit formation in the real world. *Eur. J. Soc. Psychol.* **40**, 998–1009 (2010).

256. Bennett, D. A., Schneider, J. A., Bienias, J. L., Evans, D. A. & Wilson, R. S. Mild cognitive impairment is related to Alzheimer disease pathology and cerebral infarctions. *Neurology* **64**, 834–841 (2005).

257. Kim, J., Basak, J. M. & Holtzman, D. M. The Role of Apolipoprotein E in Alzheimer's Disease. *Neuron* vol. 63 287–303 (2009).

258. Akiyama, H. *et al.* Inflammation and Alzheimer's disease. *Neurobiology of Aging* vol. 21 383–421 (2000).

259. Tang, W. J. Targeting Insulin-Degrading Enzyme to Treat Type 2 Diabetes Mellitus. *Trends in Endocrinology and Metabolism* vol. 27 24–34 (2016).

260. Sharma, S. K. *et al.* Insulin-degrading enzyme prevents α-synuclein fibril formation in a nonproteolytical manner. *Sci. Rep.* **5**, (2015).

261. Sugar's 'tipping point' link to Alzheimer's disease revealed -- ScienceDaily. https://www.sciencedaily.com/releases/2017/02/170223124253.htm.

262. Zilliox, L. A., Chadrasekaran, K., Kwan, J. Y. & Russell, J. W. Diabetes and Cognitive Impairment. *Current Diabetes Reports* vol. 16 1–11 (2016).

263. Institute of Medicine Food Forum (US). Nutrition Concerns for Aging Populations. (2010).

264. Chai, G. S. *et al.* Betaine attenuates Alzheimer-like pathological changes and memory deficits induced by homocysteine. *J. Neurochem.* **124**, 388–396 (2013).

265. Bathina, S. & Das, U. N. Brain-derived neurotrophic factor and its clinical Implications. *Archives of Medical Science* vol. 11 1164–1178 (2015).

266. Beilharz, J. E., Maniam, J. & Morris, M. J. Diet-induced cognitive deficits: The role of fat and sugar, potential mechanisms and nutritional interventions. *Nutrients* vol. 7 6719–6738 (2015).

267. Gomez-Pinilla, F. The influences of diet and exercise on mental health through hormesis. *Ageing Research Reviews* vol. 7 49–62 (2008).

268. Kivipelto, M. *et al.* Obesity and vascular risk factors at midlife and the risk of dementia and Alzheimer disease. *Arch. Neurol.* **62**, 1556–1560 (2005).

269. Chang, L. W. Neurotoxic effects of mercury-A review. *Environmental Research* vol. 14 329–373 (1977).

270. Chung, R. T. M. Detoxification effects of phytonutrients against environmental toxicants and sharing of clinical experience on practical applications. *Environ. Sci. Pollut. Res.* **24**, 8946–8956 (2017).

271. Gomm, W. *et al.* Association of proton pump inhibitors with risk of dementia: A pharmacoepidemiological claims data analysis. *JAMA Neurol.* **73**, 410–416 (2016).

272. Woolley, C. S. Acute Effects of Estrogen on Neuronal Physiology. *Annu. Rev. Pharmacol. Toxicol.* **47**, 657–680 (2007).

273. McEwen, B. *et al.* Tracking the estrogen receptor in neurons: Implications for estrogen-induced synapse formation. *Proc. Natl. Acad. Sci. U. S. A.* **98**, 7093–7100 (2001).

274. Savolainen-Peltonen, H. *et al.* Use of postmenopausal hormone therapy and risk of Alzheimer's disease in Finland: Nationwide case-control study. *BMJ* **364**, (2019).

275. Wei, P., Liu, M., Chen, Y. & Chen, D. C. Systematic review of soy isoflavone supplements on osteoporosis in women. *Asian Pac. J. Trop. Med.* **5**, 243–248 (2012).

276. Adlercreutz, H. & Mazur, W. Phyto-oestrogens and Western diseases. *Annals of Medicine* vol. 29 95–120 (1997).

277. Dominy, S. S. *et al.* Porphyromonas gingivalis in Alzheimer's disease brains: Evidence for disease causation and treatment with small-molecule inhibitors. *Sci. Adv.* **5**, (2019).

278. Ilievski, V. *et al.* Chronic oral application of a periodontal pathogen results in brain inflammation, neurodegeneration and amyloid beta production in wild type mice. *PLoS One* **13**, (2018).

279. Rogers, R. L., Meyer, J. S. & Mortel, K. F. After Reaching Retirement Age Physical Activity Sustains Cerebral Perfusion and Cognition. *J. Am. Geriatr. Soc.* **38**, 123–128 (1990).

280. Erickson, K. I., Gildengers, A. G. & Butters, M. A. Physical activity and brain plasticity in late adulthood. *Dialogues Clin. Neurosci.* **15**, 99–108 (2013).

281. Steele, C. J., Bailey, J. A., Zatorre, R. J. & Penhune, V. B. Early musical training and white-matter plasticity in the corpus callosum: Evidence for a sensitive period. *J. Neurosci.* **33**, 1282–1290 (2013).

282. Brain Scientists Identify Links between Arts, Learning | SharpBrains. https://sharpbrains.com/blog/2009/05/24/brain-scientists-identify-links-between-arts-learning/.

283. Jessen, N. A., Munk, A. S. F., Lundgaard, I. & Nedergaard, M. The Glymphatic System: A Beginner's Guide. *Neurochem. Res.* **40**, 2583–2599 (2015).

284. Bredesen, D. E. Reversal of cognitive decline: A novel therapeutic program. *Aging* vol. 6 707–717 (2014).

285. Nash, D. T. & Slutzky, A. R. Gluten Sensitivity: New Epidemic or New Myth? *Baylor Univ. Med. Cent. Proc.* **27**, 377–378 (2014).

286. Calabrese, C. *et al.* Effects of a standardized Bacopa monnieri extract on cognitive performance, anxiety, and depression in the elderly: A randomized, double-blind, placebo-controlled trial. *J. Altern. Complement. Med.* **14**, 707–713 (2008).

287. Stough, C. *et al.* Examining the nootropic effects of a special extract of Bacopa monniera on human cognitive functioning: 90 Day double-blind placebo-controlled randomized trial. *Phyther. Res.* **22**, 1629–1634 (2008).

288. Stough, C. *et al.* Examining the cognitive effects of a special extract of Bacopa monniera (CDRI08: Keenmnd): a review of ten years of research at Swinburne University. *J. Pharm. Pharm. Sci.* **16**, 254–8 (2013).

289. Stough, C. *et al.* The chronic effects of an extract of Bacopa monniera (Brahmi) on cognitive function in healthy human subjects. *Psychopharmacology (Berl).* **156**, 481–484 (2001).

290. Kongkeaw, C., Dilokthornsakul, P., Thanarangsarit, P., Limpeanchob, N. & Norman Scholfield, C. Meta-analysis of randomized controlled trials on cognitive effects of Bacopa monnieri extract. *J. Ethnopharmacol.* **151**, 528–535 (2014).

291. Vollala, V. R., Upadhya, S. & Nayak, S. Enhanced dendritic arborization of amygdala neurons during growth spurt periods in rats orally intubated with Bacopa monniera extract. *Anat. Sci. Int.* **86**, 179–188 (2011).

292. Nouchi, R. *et al.* Brain Training Game Boosts Executive Functions, Working Memory and Processing Speed in the Young Adults: A Randomized Controlled Trial. *PLoS One* **8**, e55518 (2013).

293. Holdcroft, A. Gender bias in research: How does it affect evidence based medicine? *Journal of the Royal Society of Medicine* vol. 100 2–3 (2007).

294. Michael F. Holick, M. P. Does Vitamin D Have a Role in Cancer Prevention? (2019).

295. Misra, S. Randomized double blind placebo control studies, the "Gold Standard" in intervention based studies. *Indian J. Sex. Transm. Dis. AIDS* **33**, 131 (2012).

296. Randomized, double-blind, placebo-controlled clinical trial of the efficacy of treatment with zinc or vitamin A in infants and young children with severe acute lower respiratory infection | The American Journal of Clinical Nutrition | Oxford Academic. https://academic.oup.com/ajcn/article/79/3/430/4690137.

297. Vitamins K1 and K2: The Emerging Group of Vitamins Required for Human Health. https://www.ncbi.nlm.nih.gov/pmc/articles/PMC5494092/.

298. Hewlings, S. & Kalman, D. Curcumin: A Review of Its' Effects on Human Health. *Foods* **6**, 92 (2017).

299. Olthof, M. R., van Vliet, T., Boelsma, E. & Verhoef, P. Low Dose Betaine Supplementation Leads to Immediate and Long Term Lowering of Plasma Homocysteine in Healthy Men and Women. *J. Nutr.* **133**, 4135–4138 (2003).

300. Fish: Friend or Foe? | The Nutrition Source | Harvard T.H. Chan School of Public Health. https://www.hsph.harvard.edu/nutritionsource/fish/#1.

301. Knoops, K. T. B. *et al.* Mediterranean diet, lifestyle factors, and 10-year mortality in elderly European men and women: The HALE project. *J. Am. Med. Assoc.* **292**, 1433–1439 (2004).

302. The truth about fats: the good, the bad, and the in-between - Harvard Health. https://www.health.harvard.edu/staying-healthy/the-truth-about-fats-bad-and-good.

303. Dyall, S. C. Long-chain omega-3 fatty acids and the brain: a review of the independent and shared effects of EPA, DPA and DHA. *Front. Aging Neurosci.* **7**, (2015).

304. Harris, W. S. *et al.* Omega-6 fatty acids and risk for cardiovascular disease: A science advisory from the American Heart Association subcommittee of the council on nutrition. *Circulation* **119**, 902–907 (2009).

305. Estruch, R. *et al.* Primary Prevention of Cardiovascular Disease with a Mediterranean Diet. *N. Engl. J. Med.* **368**, 1279–1290 (2013).

306. Olive oil, genes and health - NHS. https://www.nhs.uk/news/genetics-and-stem-cells/olive-oil-genes-and-health/.

307. Beilharz, J. E., Maniam, J. & Morris, M. J. Short-term exposure to a diet high in fat and sugar, or liquid sugar, selectively impairs hippocampal-dependent memory, with differential impacts on inflammation. *Behav. Brain Res.* **306**, 1–7 (2016).

308. Yu, X., Bao, Z., Zou, J. & Dong, J. Coffee consumption and risk of cancers: A meta-analysis of cohort studies. *BMC Cancer* **11**, 96 (2011).

309. Chowdhury, R. *et al.* Association of dietary, circulating, and supplement fatty acids with coronary risk: A systematic review and meta-analysis. *Ann. Intern. Med.* **160**, 398–406 (2014).

310. Gepner, Y. *et al.* Effects of initiating moderate alcohol intake on cardiometabolic risk in adults with type 2 diabetes: A 2-year randomized, controlled trial. *Ann. Intern. Med.* **163**, 569–579 (2015).

311. Haseeb, S., Alexander, B. & Baranchuk, A. Wine and Cardiovascular Health. *Circulation* **136**, 1434–1448 (2017).

312. Griswold, M. G. *et al.* Alcohol use and burden for 195 countries and territories, 1990-2016: A systematic analysis for the Global Burden of Disease Study 2016. *Lancet* **392**, 1015–1035 (2018).

313. Queipo-Ortuño, M. I. *et al.* Influence of red wine polyphenols and ethanol on the gut microbiota ecology and biochemical biomarkers. *Am. J. Clin. Nutr.* **95**, 1323–1334 (2012).

314. Crowley, J., Ball, L. & Hiddink, G. J. Nutrition in medical education: a systematic review. *Lancet Planet. Heal.* **3**, e379–e389 (2019).

315. Christen, W. G., Gaziano, J. M. & Hennekens, C. H. Design of Physicians' Health Study II—A Randomized Trial of Beta-Carotene, Vitamins E and C, and Multivitamins, in Prevention of Cancer, Cardiovascular Disease, and Eye Disease, and Review of Results of Completed Trials. *Ann. Epidemiol.* **10**, 125–134 (2000).

316. Fortmann, S. P., Burda, B. U., Senger, C. A., Lin, J. S. & Whitlock, E. P. Vitamin and Mineral Supplements in the Primary Prevention of Cardiovascular Disease and Cancer: An Updated Systematic Evidence Review for the U.S.

Preventive Services Task Force. *Ann. Intern. Med.* **159**, 824–834 (2013).

317.   Chew, E. Y. *et al.* Ten-year follow-up of age-related macular degeneration in the age-related eye disease study: AREDS report No. 36. *JAMA Ophthalmol.* **132**, 272–277 (2014).

318.   Marles, R. J. Mineral nutrient composition of vegetables, fruits and grains: The context of reports of apparent historical declines. *Journal of Food Composition and Analysis* vol. 56 93–103 (2017).

319.   Bordoni, A. *et al.* Dairy products and inflammation: A review of the clinical evidence. *Critical Reviews in Food Science and Nutrition* vol. 57 2497–2525 (2017).

320.   Chart Shows What the World's Land Is Used For ... and It Explains Exactly Why So Many People Are Going Hungry - One Green Planet. https://www.onegreenplanet.org/news/chart-shows-worlds-land-used/.

321.   Poore, J. & Nemecek, T. Reducing food's environmental impacts through producers and consumers. http://science.sciencemag.org/.

322.   Look inside the home of the future. https://www.telegraph.co.uk/wellbeing/future-health/home-of-the-future/.

323.   This High-Tech Toilet Seat Can Detect Heart Failure. https://futurism.com/neoscope/toilet-seat-heart-failure.

324.   Here's how smart toilets of the future could protect your health. https://www.nbcnews.com/mach/science/here-s-how-smart-toilets-future-could-protect-your-health-ncna961656.

325.   Dias, D. & Cunha, J. P. S. Wearable health devices—vital sign monitoring, systems and technologies. *Sensors (Switzerland)* vol. 18 (2018).

326.   This new Bluetooth-connected toothbrush brings a dentist into your bathroom - The Verge. https://www.theverge.com/circuitbreaker/2016/6/9/11877586/phillips-sonicare-connected-toothbrush-dentist-app.

327.   The Top 8 Things to Know About Anti-Aging Research Right Now - leapsmag. https://leapsmag.com/the-top-8-things-to-know-about-anti-aging-research-right-now/.

328.   Biogerontology - Research Foundation. http://bg-rf.org.uk/.

329.   Barbany, G. *et al.* Cell-free tumour DNA testing for early detection of cancer – a potential future tool. *Journal of Internal Medicine* vol. 286 118–136 (2019).

330.   Moving Towards Individualized Medicine For All | The Scientist Magazine. https://www.the-scientist.com/editorial/monogrammed-medicine-66089.

- APPENDIX 3 -

Index